Springer Proceedings in Mathematics & Statistics

Volume 266

Springer Proceedings in Mathematics & Statistics

This book series features volumes composed of selected contributions from workshops and conferences in all areas of current research in mathematics and statistics, including operation research and optimization. In addition to an overall evaluation of the interest, scientific quality, and timeliness of each proposal at the hands of the publisher, individual contributions are all refereed to the high quality standards of leading journals in the field. Thus, this series provides the research community with well-edited, authoritative reports on developments in the most exciting areas of mathematical and statistical research today.

More information about this series at http://www.springer.com/series/10533

Victor G. Kac · Peter J. Olver
Pavel Winternitz · Teoman Özer
Editors

Symmetries, Differential Equations and Applications

SDEA-III, İstanbul, Turkey, August 2017

Springer

Editors
Victor G. Kac
Department of Mathematics
Massachusetts Institute of Technology
Cambridge, MA, USA

Pavel Winternitz
Department of Mathematics and Statistics
Université de Montréal
Montréal, QC, Canada

Peter J. Olver
School of Mathematics
University of Minnesota
Minneapolis, MN, USA

Teoman Özer
Division of Mechanics,
Faculty of Civil Engineering
İstanbul Technical University
Maslak, İstanbul, Turkey

ISSN 2194-1009 ISSN 2194-1017 (electronic)
Springer Proceedings in Mathematics & Statistics
ISBN 978-3-030-01375-2 ISBN 978-3-030-01376-9 (eBook)
https://doi.org/10.1007/978-3-030-01376-9

Library of Congress Control Number: 2018957043

Mathematics Subject Classification (2010): 22E70, 35B06, 34A05, 70H33, 70S10, 35R09, 45J05, 81R30, 34C14, 53A55, 53B25, 58K50

This Springer imprint is published by the registered company Springer Nature Switzerland AG
The registered company address is: Gewerbestrasse 11, 6330 Cham, Switzerland

Preface

This book consists of the selected, peer-reviewed, and revised papers from the **III. International Conference on Symmetries, Differential Equations and Applications (SDEA-III,** www.sdea3.org), which was held from August 14 to 17, 2017, in İstanbul Technical University, İstanbul, Turkey.

The first SDEA-I was held in Johannesburg, South Africa, in 2012, dedicated to the Bicentenary of Evariste Galois. The second SDEA-II was held in İslamabad, Pakistan, in 2014, dedicated to James Clerk Maxwell's Theory of Electromagnetism.

SDEA-III was the third conference in the series and dedicated to the Centenary of Noether's Theorem proven by the prominent German mathematician Emmy Noether. The main aim of the conference was to concentrate on many of the recent important advances in the applications of Lie groups, including a wide area of topics in interdisciplinary studies ranging from mathematical physics to financial mathematics. The topics discussed in SDEA-III included Lie theory and symmetry methods in differential equations, Lie algebras and Lie pseudogroups, supersymmetry and super-integrability, representation theory of Lie algebras, classification problems, conservation laws, and geometrical methods.

We are convinced that SDEA-III was a very successful conference that provided a productive forum for academic researchers, both junior and senior, and students to discuss and share the latest developments in the theory and applications of Lie symmetry groups. The conference included 18 parallel sessions with 76 presentations, and 66 speakers, which included the 13 invited speakers listed below, from 16 different countries, as well as 17 poster representations during the 4 days it ran.

The invited speakers of SDEA-III were Prof. Mohammad Akbar (*University of Texas at Dallas, USA*), Prof. Alexei Cheviakov (*University of Saskatchewan, Canada*), Prof. Metin Gürses (*Bilkent University, Turkey*), Prof. Victor G. Kac (*MIT, USA*), Prof. Varga Kalantarov (*Koç University, Turkey*), Prof. Masood Khalique (*North-West University, South Africa*), Prof. Sergey Meleshko (*Suranaree University of Technology, Thailand*), Prof. Maria Concepcion Muriel (*University of Cádiz, Spain*), Prof. Maria Clara Nucci (*University of Perugia, Italy*), Prof. Peter J. Olver (*University of Minnesota, USA*), Prof. Kamal Soltanov (*Hacettepe University,*

Turkey), Prof. Greg Reid (*University of Western Ontario, Canada*), and Prof. Alexandre Vinogradov (*Levi-Civita Institute, Italy*).

The conference sponsors were İstanbul Metropolitan Municipality, Sarıyer Municipality, İstanbul Technical University, International Mathematical Union, Turkish Airlines, and The Scientific and Technological Research Council of Turkey. SDEA-III was organized in cooperation with SIAM, the Society for Industrial and Applied Mathematics.

As the editors, we personally wish to express our gratitude to the authors of the papers in this book, and to all participants for their contributions in this conference. We hope you will enjoy reading it and find its contributions of interest.

Cambridge, USA	Victor G. Kac
Minneapolis, USA	Peter J. Olver
İstanbul, Turkey	Teoman Özer
Montréal, Canada	Pavel Winternitz

Contents

Normal Forms for Submanifolds Under Group Actions

Peter J. Olver

Abstract We describe computational algorithms for constructing the explicit power series expansions for normal forms of submanifolds under transformation groups. The procedure used to derive the coefficients relies on the recurrence formulae for differential invariants provided by the method of equivariant moving frames.

Keywords Lie group · Submanifold · Normal form · Moving frame · Differential invariant · Curvature · Recurrence formula · Invariantization

1 Introduction

The equivariant method of moving frames, introduced in [1], provides a powerful computational tool for investigating the equivalence and symmetry properties of submanifolds under general Lie group actions (and, more generally, infinite-dimensional Lie pseudo-groups, [2, 3]), and determining the required differential invariants. The main new tool is the recurrence relations, which completely prescribe the structure of the non-commutative differential algebra they generate through the process of invariant differentiation. Remarkably, these relations and the consequent differential algebraic structure can be completely and straightforwardly constructed, requiring only basic linear algebra, and can thus be readily implemented in any modern computer algebra system, including MATHEMATICA, MAPLE, and SAGE.

A simple example is provided by the Euclidean geometry of space curves $C \subset \mathbb{R}^3$, under the action of the group of rigid motions — translations and rotations. The fundamental differential invariants are the curvature and torsion of the space curve, and the invariant differential operator is differentiation with respect to arc length. As a consequence, every Euclidean differential invariant can be expressed as a function of curvature, torsion, and their successive arc-length derivatives.

P. J. Olver (✉)
School of Mathematics, University of Minnesota, Minneapolis, MN 55455, USA
e-mail: olver@umn.edu

V. G. Kac et al. (eds.), *Symmetries, Differential Equations and Applications*,
Springer Proceedings in Mathematics & Statistics 266,
https://doi.org/10.1007/978-3-030-01376-9_1

The normalization procedure underlying the construction of a moving frame is equivalent to the specification of a "normal form" for submanifolds under the group action. Roughly, to construct a normal form, one uses the group transformations to simplify, as much as possible, the Taylor expansion of the submanifold at a given point. The result will be called a *normal form* for the submanifold at the point, also known as the *Monge* or *Monge–Taylor form*, [4, 5]. As we note below, this simplification is exactly the same as the choice of cross-section to the prolonged group orbits, which is the first step in the equivariant moving frame construction. Once a normal form has been specified, the non-constant coefficients in the resulting Taylor series expansion form a complete system of differential invariants, known, in the equivariant approach, as the fundamental normalized differential invariants.

The purpose of the present note is to explain, in simplified form, the moving frame algorithms and recurrence formulae, and how they can be used to construct the normal form expansion of a submanifold in terms of the fundamental differential invariants and their invariant derivatives. While direct calculations can be very tedious, if not impossible due to the limitations of current computer algebra software and hardware, the recurrence formulae provide a simple, straightforward route to the desired formulae. In this paper, we describe this calculus, first in the simplest context of plane curves, and then for general submanifolds under Lie group actions. The results are illustrated by a few basic examples of geometric and imaging importance.

2 Plane Curves

For simplicity, we first describe the normal form construction in the its most basic manifestation: plane curves under "ordinary" group actions. The general version can be found below in Sect. 3.

Throughout this section, $C \subset M = \mathbb{R}^2$ will denote a regular, smooth[1] (C^∞) plane curve. We use $z = (x, u)$ as local coordinates on M, and $t \in I \subset \mathbb{R}$ as a curve parameter, so that C is the image of the function $z(t) = (x(t), u(t))$ for t in the interval I. Regularity requires that the curve's tangent vector is nowhere vanishing[2]: $dz/dt = (x_t, u_t) \neq \mathbf{0}$. We will identify parametrizations that have identical image curves, meaning that we allow reparametrization, including those that reverse orientation. In particular, the curve is a *graph* if it is parametrized by the horizontal coordinate x, so that $z(x) = (x, u(x))$ for $x \in I \subset \mathbb{R}$. Locally, in a neighborhood of $z_0 = (x_0, u_0) \in C$, a curve can be parametrized uniquely as a graph if and only if it intersects the vertical fiber $\{\, x = x_0 \,\}$ transversally, meaning that its tangent vector at z_0 is not vertical, i.e., $x_t \neq 0$ there.

[1]One can apply the construction to curves of class C^n provided n is sufficiently large that all derivatives indicated are continuous.

[2]Subscripts on dependent variables indicate derivatives.

Given a graph defined by the function $u(x)$, we will identify its Taylor polynomial of order n at a point $z_0 = (x_0, u_0) = (x_0, u(x_0)) \in C$, namely,[3]

$$u(x_0) + u_x(x_0)\,(x - x_0) + \frac{1}{2}\,u_{xx}(x_0)\,(x - x_0)^2 + \cdots + \frac{1}{n!}\,u_n(x_0)\,(x - x_0)^n,$$

with the nth order *jet* of the curve at the point z_0. Note that the n jet is uniquely prescribed by the derivatives of order $\leq n$ at the point in question. Thus, the space of nth order transverse[4] curve jets, denoted J^n, can be identified with $\mathbb{R}^{n+2}$, with coordinates

$$z^{(n)} = (x, u, u_x, u_{xx}, \ldots, u_n). \tag{1}$$

The n-jet of the graph $C = \{(x, u(x))\}$ at the point $z_0 = (x_0, u(x_0)) \in C$ is thereby identified with the $(n + 2)$–tuple

$$\mathrm{j}_n C|_{z_0} = \big(x_0, u(x_0), u_x(x_0), u_{xx}(x_0), \ldots, u_n(x_0)\big) \in \mathrm{J}^n. \tag{2}$$

One can straightforwardly derive, via implicit differentiation, expressions for the curve jet components (2) in terms of a general parametrization $z(t) = (x(t), u(t))$, writing the nth order jet coordinate u_n as an explicit rational function of the derivatives, of order $\leq n$, of $x(t)$, $u(t)$. For example,

$$u_x = D_x u = \frac{u_t}{x_t}, \qquad u_{xx} = D_x u_x = \frac{1}{x_t} D_t \left(\frac{u_t}{x_t}\right) = \frac{x_t u_{tt} - u_t x_{tt}}{x_t^3} \quad \cdots, \tag{3}$$

with the higher order expressions obtained by iteratively applying the implicit total derivative operator

$$D_x = \frac{1}{x_t} D_t. \tag{4}$$

By a *differential function*, we mean a (locally defined) real-valued function on the jet space, $F \colon \mathrm{J}^n \to \mathbb{R}$, and so, in coordinates, taking form

$$F(z^{(n)}) = F(x, u, u_x, u_{xx}, \ldots, u_n).$$

To us, the most important differential functions are the differential invariants, e.g., curvature, torsion, and the like. Note that one can use the parametric differentiation formulae (3) to re-express any differential function in terms of a general curve parametrization.

Let G be an r-dimensional Lie group acting on $M = \mathbb{R}^2$. There is an induced action of G on curves, with $g \in G$ mapping the curve C parametrized by $z(t)$ to the image curve $\widetilde{C} = g \cdot C$ parametrized by $\widetilde{z}(t) = g \cdot z(t)$. Two curves $C, \widetilde{C} \subset M$ are said to be *equivalent* if there exists a group element $g \in G$ such that $\widetilde{C} = g \cdot C$.

---

[3] In this section, u_n represents the nth order derivative of u with respect to x.

[4] See [6] for the extended jet bundle construction, that includes non-transverse curves.

Again, we allow reparametrization in our identification of curves. In practice, we are primarily interested in local equivalence, in the neighborhood of corresponding points on the two curves.

The action of G on curves induces an action on their jets. In other words, given a jet $z_0^{(n)} \in \mathrm{J}^n|_{z_0}$, let C be any transverse curve whose jet at $z_0 \in C$ coincides with $z_0^{(n)}$ at the point $z_0 \in C$. Then $g \cdot z_0^{(n)}$ is equal to the n-jet of the image curve $\widetilde{C} = g \cdot C$ at the image point $\widetilde{z}_0 = g \cdot z_0$. If the image curve is not transverse, the action is not defined in the ordinary jet space (although it is defined on the extended jet bundle, cf. [6]), meaning that the prolonged group action on J^n is, in general, only a local action even if the action on M is global. The explicit formulae for the prolonged action of a transformation group are obtained by implicit differentiation, [1, 6].

A *differential invariant* of *order n* is a differential function $I(z^{(n)})$ that is unaffected by the prolonged group action, i.e., $I(g \cdot z^{(n)}) = I(z^{(n)})$ for all $g \in G$ and all $z^{(n)} \in \mathrm{J}^n$, where defined. Clearly, equivalent curves have identical differential invariants, although, of course, their explicit formulae in terms of the curves' individual parametrizations may vary. The Cartan solution to the equivalence problem, [7], is based on the functional identities, or *syzygies*, among the differential invariants which are used to parametrize the associated signature. (In the case of curves in Euclidean space, the signature curve was introduced earlier by Bruce and Giblin, [4], under the name "Monge-Taylor map".) See, for example, [8–13] for various applications of the differential invariant signature to object recognition in digital images.

In its simplest incarnation, a *cross-section* to the prolonged group action is a fixed jet $z_0^{(n)} \in \mathrm{J}^n$ with the property that for any (nearby) curve C and point $z \in C$ there is a *unique* group element $g \in G$ such that

$$g \cdot \left(\mathrm{j}_n C|_z \right) = \mathrm{j}_n (g \cdot C)|_{z_0} = z_0^{(n)}, \tag{5}$$

meaning that the group element maps the curve jet at z to the fixed cross-section jet. In particular $g \cdot z = z_0$. A straightforward chain rule argument demonstrates that the group element satisfying (5) depends only of the n-jet $z^{(n)} = \mathrm{j}_n C|_z$ of the curve at the point z. In view of uniqueness, we write $g = \rho(z^{(n)})$, whereby (5) is equivalent to the equation

$$\rho(z^{(n)}) \cdot z^{(n)} = z_0^{(n)}. \tag{6}$$

In the language of [1], the map[5] $\rho \colon \mathrm{J}^n \to G$ defines a (right) *moving frame* of *order n*, and, as can be easily proved, satisfies the *right equivariance* rule

$$\rho(g \cdot z^{(n)}) = \rho(z^{(n)}) \cdot g^{-1}, \tag{7}$$

where the dot on the left hand side indicates the prolonged group action on J^n, while the dot on the right hand side represents group multiplication. Occasionally, formulae are more simply written in terms of the corresponding *left equivariant moving frame*, which is merely the group inverse of the right moving frame:

[5] Typically ρ is only defined on an open subset of the jet space.

$$\widetilde{\rho}(z^{(n)}) = \rho(z^{(n)})^{-1}, \quad \text{satisfying the } \textit{left equivariance} \text{ rule } \widetilde{\rho}(g \cdot z^{(n)}) = g \cdot \widetilde{\rho}(z^{(n)}), \tag{8}$$

and mapping the normal form jet to the curve jet: $\widetilde{\rho}(z^{(n)}) \cdot z_0^{(n)} = z^{(n)}$.

Given a choice of cross-section, a curve C_0 is said to be in *normal form* if $z_0 \in C_0$ and its n-jet at z_0 coincides with the fixed cross-section jet: $\mathrm{j}_n C_0|_{z_0} = z_0^{(n)}$. Thus, given

$$z_0^{(n)} = (x_0, u_0 = c_0, c_1, \ldots, c_n),$$

any *normal form* curve, parametrized as the graph of the function $u_0(x)$, has Taylor expansion

$$\begin{aligned} u_0(x) = {} & c_0 + c_1\,(x - x_0) + \frac{1}{2}\,c_2\,(x - x_0)^2 + \cdots + \frac{1}{n!}\,c_n\,(x - x_0)^n \\ & + \frac{1}{(n+1)!}\,u_{n+1}(x_0)\,(x - x_0)^{n+1} + \frac{1}{(n+2)!}\,u_{n+2}(x_0)\,(x - x_0)^{n+2} + \cdots, \end{aligned} \tag{9}$$

at $x = x_0$, whose first $n + 1$ coefficients are fixed by the choice of cross-section jet, whereas the values of those of order $\geq n + 1$ depend upon the particularities of the curve C_0.

Remark: Existence of a cross-section of the above type is equivalent to the transitivity and freeness[6] of the prolonged group action on an open subset of J^n. If the Lie group G has dimension r, then this requires $n = r - 2$. A planar group action that admits a cross-section in the above sense is known as *ordinary*, [7]. The only non-ordinary group actions on $\mathbb{R}^2$ are intransitive actions and those whose prolongations exhibit *pseudo-stabilization*, meaning that they act intransitively but not freely on some jet space. All "standard" transitive group actions arising in geometry and image processing are ordinary. Moreover, non-ordinary actions can be readily handled by the general moving frame construction described in the following section.

Applying the moving frame group element $g = \rho(z^{(n)})$ to the curve C produces the *normal form curve* $C_0 = g \cdot C = \rho(z^{(n)}) \cdot C$ associated with the point $z \in C$, that satisfies the normal form constraint $\mathrm{j}_n C_0 = z_0^{(n)}$. Clearly, two curves are locally equivalent if and only if they have identical normal forms at the matching points. Consequently, each Taylor coefficient of the normal form curve at the point z_0, when expressed as a function of the original curve jet, defines a differential invariant. In other words, for any k,

$$z_0^{(k)} = \mathrm{j}_k C_0|_{z_0} = \mathrm{j}_k\big(\rho(z^{(n)}) \cdot C\big)|_{z_0} = \rho(z^{(n)}) \cdot \big(\mathrm{j}_k C|_z\big) = \rho(z^{(n)}) \cdot z^{(k)} = I^{(k)}(z^{(k)}), \tag{10}$$

defines a vector-valued differential invariant: $I^{(k)}(g \cdot z^{(k)}) = I^{(k)}(z^{(k)})$ for all $g \in G$ where defined, whose individual components provide $k + 2$ scalar-valued differential invariants[7]:

---

[6]The action of G is *free* at $z^{(n)} \in \mathrm{J}^n$ if the only group element that fixes $z^{(n)}$ is the identity, i.e., $g \cdot z^{(n)} = z^{(n)}$ if and only if $g = e$.

[7]When $k \leq n$, then $I^{(k)} = \big(x_0, c_0, c_1, \ldots, c_k\big)$ is constant.

$$I^{(k)}(z^{(k)}) = \big(H(z^{(k)}), I_0(z^{(k)}), I_1(z^{(k)}), \ldots, I_k(z^{(k)}) \big) \\ = \big(x_0, c_0, c_1, \ldots, c_n, I_{n+1}(z^{(n+1)}), \ldots, I_k(z^{(k)}) \big). \tag{11}$$

As a result, the normal form Taylor expansion (9) is

$$u_0(x) = c_0 + c_1\,(x - x_0) + \frac{1}{2}\,c_2\,(x - x_0)^2 + \cdots + \frac{1}{n!}\,c_n\,(x - x_0)^n + \\ + \frac{1}{(n+1)!}\,I_{n+1}(z^{(n+1)})\,(x - x_0)^{n+1} + \cdots + \frac{1}{k!}\,I_k(z^{(k)})\,(x - x_0)^k + \cdots, \tag{12}$$

We will call $I_j(z^{(j)})$ the jth order *normalized differential invariant*; note that its value is independent of the choice of $k \geq j$ in (11); indeed, it would be convenient to set $k = \infty$ and work with Taylor series (infinite jets) throughout. Of course, the first $n + 2$ of these, $H, I_0, \ldots, I_n$, are constant, since they equal the corresponding Taylor coefficient (11) of the cross-section jet: $I_j(z^{(n)}) = c_j$.

According to [1], the non-phantom or *fundamental normalized differential invariants* of order $> n$, namely $I_{n+1}(z^{(n+1)}), I_{n+2}(z^{(n+2)}), \ldots$, form a complete system of differential invariants for the action of G on curves, meaning that, locally, any other differential invariant can be written, uniquely, as a function thereof. Indeed, the *Replacement Rule* states that if $J(z^{(k)}) = J(x, u, u_x, \ldots, u_k)$ is any differential invariant of order[8] $k > n$, then, replacing each of its arguments by the corresponding normalized invariant,

$$J(z^{(k)}) = J\big(x_0, c_0, \ldots, c_n, I_{n+1}(z^{(n+1)}), \ldots, I_k(z^{(k)}) \big) \tag{13}$$

gives an explicit formula for J in terms of the fundamental normalized invariants. In symbolic computation terminology, [14], (13) is a *rewrite rule* expressing any differential invariant in terms of the fundamental generators.

Further, the moving frame map induces a process of *invariantization*, denoted by ι, that associates a differential invariant with any differential function. Namely, if $F(z^{(k)})$ is any function of the curve jets, then its invariantization $J(z^{(k)}) = \iota\big[\, F(z^{(k)}) \,\big]$ is the unique differential invariant that agrees with the value of F on the normal form prescribed by the cross-section: $J(z_0^{(k)}) = F(z_0^{(k)})$. Note that invariantization respects all algebraic operations — but *not* differentiation, which is the point of the recurrence formulae derived below. It is not hard to see that the invariantization process is readily implemented by substituting each jet coordinate appearing in the argument of F by the corresponding normalized differential invariant:

$$\iota\big[\, F(z^{(k)}) \,\big] = F\big(x_0, c_0, \ldots, c_n, I_{n+1}(z^{(n+1)}), \ldots, I_k(z^{(k)}) \big). \tag{14}$$

Furthermore, invariantization does not affect differential invariants: $\iota\big[\, J(z^{(k)}) \,\big] = J(z^{(k)})$ and hence, comparison with (14) immediately establishes the Replacement Rule (13).

[8]Since G acts transitively on J^n, any differential invariant of order $\leq n$ is necessarily constant, and still satisfies the Replacement Rule.

Example 1 Plane curves under orientation-preserving rigid motions: In this example, $G = \mathrm{SE}(2)$ is the special Euclidean group, consisting of translations and rotations of $\mathbb{R}^2$:

$$x \mapsto x\cos\phi - u\sin\phi + a, \qquad u \mapsto x\sin\phi + u\cos\phi + b, \qquad \begin{array}{l} a, b \in \mathbb{R}, \\ -\pi < \phi \le \pi. \end{array} \tag{15}$$

To place a plane curve in Euclidean normal form at a point $z \in C$, we first use the translations to move the base point to the origin, $x_0 = u_0 = 0$, and then rotate the translated curve so that its tangent is horizontal, whereby $u_{x,0} = 0$. The resulting *Euclidean normal form* for a plane curve has Taylor expansion

$$u_0(x) = \tfrac{1}{2}\, I_2\, x^2 + \tfrac{1}{6}\, I_3\, x^3 + \tfrac{1}{24}\, I_4\, x^4 + \cdots + \frac{1}{k\,!}\, I_k\, x^k + \cdots \tag{16}$$

at the origin. Its Taylor coefficients

$$I_k = \iota(u_k), \qquad k \ge 2, \tag{17}$$

when expressed in terms of the original curve parametrization, are the fundamental normalized differential invariants.

The preceding choice of normal form corresponds to the cross-section

$$x = u = u_x = 0, \tag{18}$$

whence the three associated phantom invariants are

$$\iota(x) = H = 0, \qquad \iota(u) = I_0 = 0, \qquad \iota(u_x) = I_1 = 0. \tag{19}$$

The resulting left moving frame[9] $\widetilde{\rho} : \mathrm{J}^1 \to \mathrm{SE}(2)$ can be identified with the classical moving frame, [15], namely, its translation component is the point $z \in C$, while the columns of the rotation matrix, $R = [\,\mathbf{t}, \mathbf{n}\,]$, are the orthonormal frame vectors based at z, that is, the unit tangent $\mathbf{t}$ and normal $\mathbf{n}$. Furthermore, by direct computation or, alternatively, by applying the moving frame construction, the lowest order normalized differential invariant

$$I_2 = \iota(u_{xx}) = \kappa = \frac{u_{xx}}{(1 + u_x^2)^{3/2}} \tag{20}$$

turns out to be the Euclidean curvature of the curve. We defer the identification of the higher order normalized invariants $I_3, I_4, \ldots$, until we have constructed the associated recurrence formulae.

---

[9]Typically, while the right moving frame (7) plays a more fundamental role and is easier to compute, in classical geometries, the corresponding left moving frame (8) includes the usual frame vectors on the submanifold, cf. [1].

Remark: In the preceding example, there remains an unresolved discrete ambiguity since we can rotate by π radians without affecting the cross-section (18). The effect is to map $u_0(x)$ to $-u_0(-x)$, and hence change the sign of the even order normalized invariants, $I_{2j} \mapsto -I_{2j}$, so that, in particular, the curvature invariant changes sign: $\kappa \mapsto -\kappa$. This can be avoided by either working with its absolute value or, if one restricts attention to closed curves, by fixing the orientation. Here, to avoid technicalities, we will ignore this final ambiguity (as is done in most treatments), referring to [16] for the full details, including the additional effects of reflections on the moving frame and differential invariants.

An alternative method of generating differential invariants is through invariant differentiation. Given a transformation group acting on plane curves, we use ds to denote the G-invariant arc-length element, or, equivalently, the invariant[10] one-form of lowest order. We remark that the invariant one-forms can also be systematically constructed through a reasonably straightforward extension of the invariantization process associated with the moving frame, and refer the reader to [1, 17] for details.

Let $\mathscr{D} = d/ds$ be the dual invariant differential operator, i.e., the arc length derivative. Invariance of the arc-length form ds implies that $\mathscr{D}$ maps a differential invariant of order k to a differential invariant of order $k+1$. In particular, starting with the (non-constant) normalized differential invariant $\kappa = I_{n+1}$ of lowest order, namely $n+1$, which we identify[11] as the G-invariant *curvature* function, its successive arc-length derivatives $\kappa_s = \mathscr{D}\kappa,\ \kappa_{ss} = \mathscr{D}^2\kappa, \ldots$, are differential invariants of respective orders $n+2, n+3, \ldots$. It is known that they also generate the algebra of differential invariants; one way of proving this assertion is by inspection of the recurrence formulae. The Replacement Rule (13) tells us that these are all functions of the normalized differential invariants; vice versa, it can be shown that the normalized differential invariants are themselves certain functions of the curvature invariant and its successive arc length derivatives. The resulting formulae

$$I_k = F_k(\kappa, \kappa_s, \ldots, \kappa_{k-n-1}), \qquad k \geq n+1, \tag{21}$$

enable one to express the coefficients of the normal form Taylor expansion (12) of a curve in terms of the curvature invariant and its arc-length derivatives. Our goal is to develop the machinery that enables one to straightforwardly compute these formulas, and hence the explicit Taylor expansion for the normal form of a curve under a group action.

While, in principle, knowing the explicit coordinate formulae for the curvature invariants enables one, e.g. via the Replacement Rule (13), to express them in terms

[10]Strictly speaking, ds is only "contact-invariant", meaning that it is not an invariant form on jet space but, rather, is invariant when restricted to curve jets, or, equivalently, is invariant modulo contact forms, [7].

[11]Identification of κ with a classical geometric quantity (Euclidean curvature, equi-affine curvature, projective curvature, etc.) requires an appropriate choice of normal form. Other choices may result in some function, e.g., a constant multiple, of the classical curvature invariant. Incidentally, the function in question can be straightforwardly found by applying the Replacement Rule (13) to the classical formula.

of the normalized invariants, and hence, by inversion, determine the desired formulae (21), in practice, for complicated group actions and higher order invariants, this can be a very cumbersome and complicated procedure that can overwhelm the abilities of even sophisticated computer algebra systems such as MATHEMATICA, MAPLE, SAGE, etc. The power of the equivariant moving frame method is that it enables one to systematically and straightforwardly derive these formulae without a priori knowledge of the explicit formulae for any of the differential invariants, or the invariant arc length derivative, or even the moving frame itself. All that is required is the formulae for the prolonged infinitesimal generators of the group action, coupled with some simple (symbolic) linear algebra!

To explain the computational algorithm, let

$$\mathbf{v}_\sigma = \xi_\sigma(x,u)\frac{\partial}{\partial x} + \varphi_\sigma(x,u)\frac{\partial}{\partial u}, \qquad \sigma = 1,\ldots,r, \tag{22}$$

be a basis for the Lie algebra of infinitesimal generators of the action of G, which are vector fields on M, [6]. Let

$$\operatorname{pr}\mathbf{v}_\sigma = \xi_\sigma(x,u)\frac{\partial}{\partial x} + \sum_{k\geq 0}\varphi_\sigma^k(x,u^{(k)})\frac{\partial}{\partial u_k}, \qquad \sigma = 1,\ldots,r, \tag{23}$$

be the corresponding infinitesimal generators of the prolonged action of G on the jet spaces, whose coefficients are explicitly determined by the well-known prolongation formula, [6]:

$$\varphi_\sigma^k(z^{(k)}) = D_x^k\big[\,\varphi_\sigma(x,u) - \xi_\sigma(x,u)\,u_x\,\big] + \xi_\sigma(x,u)\,u_{k+1}. \tag{24}$$

Here

$$D_x = \frac{\partial}{\partial x} + u_x\frac{\partial}{\partial u} + u_{xx}\frac{\partial}{\partial u_x} + u_{xxx}\frac{\partial}{\partial u_{xx}} + \cdots \tag{25}$$

is the total derivative operator, which effectively differentiates differential functions by treating u as a function of x.

The *recurrence formulae* for the differentiated invariants, [1], are

$$I_{k+1} = \mathscr{D} I_k - \sum_{\sigma=1}^{r} K_\sigma\,\iota\big[\,\varphi_\sigma^k(x,u^{(k)})\,\big], \qquad k = 0,1,2,\ldots\,, \tag{26}$$

where ι is the invariantization map (14) and $K_1,\ldots,K_r$ are certain as yet unspecified differential invariants known as the *Maurer–Cartan invariants*.[12] In particular, if one

[12]This is because they are, in fact, the coefficients of the pull-backs of the Maurer–Cartan forms via the moving frame map $\rho : \mathrm{J}^n \to G$, [1]. However, while this is essential to proving the validity of (26), from a purely practical standpoint there is no need to know this theoretical fact, or even understand what a "Maurer–Cartan form" is, since, as we will soon see, we can readily determine their explicit formulae directly from the recurrence formulae themselves.

takes $0 \leq k \leq n = r - 2$ in (26), then $I_k = c_k$ is a constant phantom invariant, and hence the first term on the right hand side of the recurrence formula is zero. Thus, the result is a system of $r - 1$ linear equations for the r Maurer–Cartan invariants in terms of the normalized differential invariants of order $\leq r - 1 = n + 1$. These are supplemented by the recurrence formula for the remaining phantom invariant $H = \iota(x) = x_0$, which takes the form

$$1 = \mathscr{D}H - \sum_{\sigma=1}^{r} K_\sigma\, \iota[\,\xi_\sigma(x,u)\,] = - \sum_{\sigma=1}^{r} K_\sigma\, \iota[\,\xi_\sigma(x,u)\,]. \tag{27}$$

It can be shown that the resulting system of r linear equations can be uniquely solved for the Maurer–Cartan invariants $K_1, \ldots, K_r$, which can thus all be expressed as certain rational functions of the curvature invariant $\kappa = I_{n+1}$. With these expressions in hand, the resulting higher order recurrence formulae (26), for $k > n$, will then iteratively provide the required formulae (21) for each I_{k+1} in terms of the arc length derivatives of κ. Let us see how this works in the context of a couple of examples.

Example 2 Return to the action of the Euclidean group on plane curves introduced in Example 1. We use

$$\mathscr{D} = D_s = \frac{1}{\sqrt{1+u_x^2}}\, D_x \tag{28}$$

to denote the invariant arc length total derivative operator.

The infinitesimal generators of the action (15) are

$$\mathbf{v}_1 = \partial_x, \qquad \mathbf{v}_2 = \partial_u, \qquad \mathbf{v}_3 = -u\,\partial_x + x\,\partial_u. \tag{29}$$

Applying the prolongation (24), the infinitesimal generators of the prolonged action of SE(2) on plane curves are

$$\begin{aligned}
\operatorname{pr}\mathbf{v}_1 &= \partial_x, \qquad\qquad \operatorname{pr}\mathbf{v}_2 = \partial_u,\\
\operatorname{pr}\mathbf{v}_3 &= -u\,\partial_x + x\,\partial_u + (1+u_x^2)\,\partial_{u_x} + 3u_x u_{xx}\,\partial_{u_{xx}} + (4u_x u_{xxx} + 3u_{xx}^2)\,\partial_{u_{xxx}} + \\
&\qquad + (5u_x u_{xxxx} + 10u_{xx}u_{xxx})\,\partial_{u_{xxxx}} + \\
&\qquad + (6u_x u_{xxxxx} + 15u_{xx}u_{xxxx} + 10u_{xxx}^2)\,\partial_{u_{xxxxx}} + \cdots.
\end{aligned} \tag{30}$$

Thus, the recurrence formulae (26), (27) for the three phantom invariants (19) are

$$\begin{aligned}
1 &= \mathscr{D}H - K_1\,\iota(1) - K_2\,\iota(0) - K_3\,\iota(-u) = -K_1,\\
0 = I_1 &= \mathscr{D}I_0 - K_1\,\iota(0) - K_2\,\iota(1) - K_3\,\iota(x) = -K_2,\\
\kappa = I_2 &= \mathscr{D}I_1 - K_1\,\iota(0) - K_2\,\iota(0) - K_3\,\iota(1+u_x^2) = -K_3,
\end{aligned}$$

and hence the Maurer–Cartan invariants are

$$K_1 = -1, \qquad K_2 = 0, \qquad K_3 = -\kappa. \tag{31}$$

Using (17), (19), these values are then iteratively substituted into the higher order recurrence formulae (26) to produce

$$\begin{aligned}
I_3 &= \mathscr{D} I_2 - K_3\, \iota(3u_x u_{xx}) = \kappa_s,\\
I_4 &= \mathscr{D} I_3 - K_3\, \iota(4u_x u_{xxx} + 3u_{xx}^2) = \kappa_{ss} + 3\kappa^3,\\
I_5 &= \mathscr{D} I_4 - K_3\, \iota(5u_x u_{xxxx} + 10u_{xx} u_{xxx}) = \kappa_{sss} + 19\kappa^2\kappa_s,\\
I_6 &= \mathscr{D} I_5 - K_3\, \iota(6u_x u_{xxxxx} + 15u_{xx} u_{xxxx} + 10u_{xxx}^2)\\
&\qquad\qquad = \kappa_{ssss} + 34\kappa^2\kappa_{ss} + 48\kappa\kappa_s^2 + 45\kappa^5,
\end{aligned} \tag{32}$$

and so on. We conclude that the explicit Taylor expansion of a curve placed in Euclidean normal form (16) is

$$\begin{aligned}
u_0(x) = \tfrac{1}{2}\kappa x^2 + \tfrac{1}{6}\kappa_s x^3 + \tfrac{1}{24}(\kappa_{ss} + 3\kappa^3)x^4 + \tfrac{1}{5!}(\kappa_{ssss} + 19\kappa^2\kappa_s)x^5 +\\
+ \tfrac{1}{6!}(\kappa_{ssss} + 34\kappa^2\kappa_{ss} + 48\kappa\kappa_s^2 + 45\kappa^5)x^6 + \cdots .
\end{aligned} \tag{33}$$

Higher order terms can be systematically constructed by continuing the above procedure. However, I do not know a general formula for the differential polynomials in κ that appear as coefficients.

Example 3 A more substantial example is provided by the geometry of equi-affine planar curves, [15], also of importance for image processing, [8]. The *equi-affine group* SA(2) acts on $M = \mathbb{R}^2$ via area-preserving affine transformations

$$g \cdot (x, u) = (\alpha x + \beta u + a, \gamma x + \delta u + b), \qquad \alpha\delta - \beta\gamma = 1. \tag{34}$$

The normalization equations

$$x = u = u_x = 0, \qquad u_{xx} = 1, \qquad u_{xxx} = 0, \tag{35}$$

define a cross-section to the prolonged action, which leads to the classical equi-affine moving frame. This normalization can be applied except at inflection points, i.e., provided the *nondegeneracy condition* $u_{xx} \neq 0$ holds. (Similar nondegeneracy conditions appear in most examples, the preceding case of Euclidean plane curves being a notable exception. At isolated inflection points one can, in principle, use the general moving frame procedure, to be presented in Sect. 3, to construct a higher order moving frame.) The cross-section (35) corresponds to the following equi-affine normal form for a non-degenerate plane curve:

$$u_0(x) = \tfrac{1}{2}x^2 + \tfrac{1}{4!} I_4 x^4 + \tfrac{1}{5!} I_5 x^5 + \tfrac{1}{6!} I_6 x^6 + \cdots . \tag{36}$$

The fundamental differential invariant is the equi-affine curvature

$$\kappa = I_4 = \iota(u_{xxxx}) = \frac{u_{xx}u_{xxxx} - \frac{5}{3}u_{xxx}^2}{u_{xx}^{8/3}}, \tag{37}$$

while

$$\mathscr{D} = \iota(D_x) = u_{xx}^{-1/3}\, D_x \tag{38}$$

is the invariant differentiation operator with respect to equi-affine arc-length. Both formulas (37), (38) can be straightforwardly found by a complete implementation of the moving frame construction, but are *not* required to perform the ensuing computations.

Our goal is to write the higher order differential invariants

$$I_k = \iota(u_k), \qquad k \geq 4, \tag{39}$$

and hence the equi-affine normal form (36), in terms of the equi-affine curvature and its arc-length derivatives. Applying (23), (24), the prolonged infinitesimal generators for the equi-affine group action (34) are

$$\begin{aligned}
\mathbf{v}_1 &= \partial_x, \\
\mathbf{v}_2 &= \partial_u, \\
\mathbf{v}_3 &= -x\,\partial_x + u\,\partial_u + 2u_x\,\partial_{u_x} + 3u_{xx}\,\partial_{u_{xx}} + 4u_{xxx}\,\partial_{u_{xxx}} + 5u_{xxxx}\,\partial_{u_{xxxx}} + \cdots, \\
\mathbf{v}_4 &= u\,\partial_x - u_x^2\,\partial_{u_x} - 3u_xu_{xx}\,\partial_{u_{xx}} - (4u_xu_{xxx} + 3u_{xx}^2)\,\partial_{u_{xxx}} - \\
&\qquad - (5u_xu_{xxxx} + 10u_{xx}u_{xxx})\,\partial_{u_{xxxx}} + \cdots, \\
\mathbf{v}_5 &= x\,\partial_u + \partial_{u_x}.
\end{aligned} \tag{40}$$

Thus, the recurrence formulae (26), (27) for the phantom invariants coming from invariantizing the cross-section coordinates (37) are

$$\begin{gathered}
1 = \mathscr{D}H - K_1 = -K_1, \quad 0 = I_1 = \mathscr{D}I_0 - K_2 = -K_2, \quad 1 = I_2 = \mathscr{D}I_1 - K_5 = -K_5, \\
0 = I_3 = \mathscr{D}I_2 - 3K_3 = -3K_3, \quad \kappa = I_4 = \mathscr{D}I_3 + 3K_4 = 3K_4,
\end{gathered}$$

and hence the Maurer–Cartan invariants are

$$K_1 = -1, \qquad K_2 = 0, \qquad K_3 = 0, \qquad K_4 = \tfrac{1}{3}I_4 = \tfrac{1}{3}\kappa, \qquad K_5 = -1. \tag{41}$$

These values are then substituted into the higher order recurrence formulae (26) to iteratively produce the desired formulae:

$$\begin{aligned}
I_5 &= \mathscr{D}I_4 = \kappa_s, \\
I_6 &= \mathscr{D}I_5 + 5I_4^2 = \kappa_{ss} + 5\kappa^2, \\
I_7 &= \mathscr{D}I_6 + 7I_4I_5 = \kappa_{sss} + 17\kappa\kappa_s, \\
I_8 &= \mathscr{D}I_7 + \tfrac{28}{3}I_4I_6 + \tfrac{35}{3}I_4^3 = \kappa_{ssss} + \tfrac{79}{3}\kappa\kappa_{ss} + 17\kappa_s^2 + \tfrac{175}{3}\kappa^3,
\end{aligned} \tag{42}$$

and so on. We conclude that the equi-affine normal form for a plane curve at a non-inflection point is given by

$$u_0(x) = \tfrac{1}{2}x^2 + \tfrac{1}{4!}\kappa x^4 + \tfrac{1}{5!}\kappa_s x^5 + \tfrac{1}{6!}(\kappa_{ss} + 5\kappa^2)x^6 + \tfrac{1}{7!}(\kappa_{sss} + 17\kappa\kappa_s)x^7 + \\ + \tfrac{1}{8!}\left(\kappa_{ssss} + \tfrac{79}{3}\kappa\kappa_{ss} + 17\kappa_s^2 + \tfrac{175}{3}\kappa^3\right)x^8 + \cdots . \tag{43}$$

Again, while they are easily found by iterating the preceding algorithm, I do not know a general explicit formula for the differential polynomials appearing in the normal form expansion (43).

3 Normal Forms for Submanifolds

We now turn to the equivariant moving frame construction, [1], that applies to completely general Lie group actions and, when suitably adapted, [2], also to infinite-dimensional Lie pseudo-group actions. Let M be an m-dimensional manifold which, since we are working locally, we identify as (an open subset of) $\mathbb{R}^m$. Given $1 \leq p < m$, there is an induced action of G on p-dimensional submanifolds $S \subset M$, and we are interested in determining when two such submanifolds are *equivalent*, meaning that there exists $g \in G$ mapping one (locally) to the other: $\widetilde{S} = g \cdot S$. As before, we are interested in the submanifold purely as a subset of M, and thus allow arbitrary reparametrizations thereof. (Although one can readily adapt the procedure to avoid or restrict allowable reparametrizations.) The solution to the equivalence problem is based on the differential invariant signature, and the moving frame method allows one to explicitly determine the fundamental differential invariants used to construct the required signature, [1, 8].

We employ coordinates $(x, u) = (x^1, \ldots, x^p, u^1, \ldots, u^q)$ on $\mathbb{R}^m$, with $p + q = m$, treating the x's as independent variables and the u's as dependent variables, whereby any p-dimensional submanifold S that is transverse to the vertical fibers $\{x = x_0\}$ can be locally identified with the graph, $S = \{(x, u(x))\}$, of a smooth vector-valued function $x \mapsto u(x)$ with components $u^\alpha(x^1, \ldots, x^p)$, $\alpha = 1, \ldots, q$. We identify the nth order Taylor expansion of $u(x)$ at a point x_0 in its domain as the *n jet* of the submanifold at the base point $z_0 = (x_0, u_0) = (x_0, u(x_0)) \in S$. The resulting nth order *jet space* J^n, for p-dimensional submanifolds, is coordinatized by the independent variables $x^1, \ldots, x^p$, the dependent variables $u^1, \ldots, u^q$, and their derivatives up to order n, which we denote by u^α_J, with $\alpha = 1, \ldots, q$, and $J = (j_1, \ldots, j_k)$ a symmetric multi-index, with $1 \leq j_k \leq p$, of order $1 \leq k = \#J \leq n$, whose entries indicate partial derivatives of u^α with respect to the x's. Thus, a point in J^n is specified by the coordinates

$$z^{(n)} = (\ldots x^i \ldots u^\alpha \ldots u^\alpha_J \ldots), \quad \text{where} \quad i = 1, \ldots, p, \quad \alpha = 1, \ldots, q, \quad \#J \leq n. \tag{44}$$

See [1, 6, 7] for details.

The action of G on submanifolds induces an action on their jets, leading to the prolonged group action on jet space. Explicit formulas are obtained by implicit differentiation. In general, a *cross-section* is a submanifold of the jet space, $\mathscr{K} \subset \mathrm{J}^n$ that has complementary dimension and is transverse to the prolonged group orbits. Moreover, we assume that, for each jet $z^{(n)} \in \mathrm{J}^n$ sufficiently close to $\mathscr{K}$, there is a *unique* group element $g = \rho(z^{(n)})$ that maps $z^{(n)}$ to the cross-section, which, as before, specifies the moving frame map[13] $\rho : \mathrm{J}^n \to G$, satisfying the right equivariance condition (7). Transversality means that no (non-zero) prolonged infinitesimal generator is tangent to the cross-section, which can be straightforwardly verified using their explicit formulas, cf. (50) below, and involves computing the rank of a certain matrix. Existence of a cross-section, and hence a moving frame, requires that the prolonged action be (locally) free and regular on an open subset of J^n, and it can be proved that, assuming the action on M is locally effective on subsets, local freeness holds at a sufficiently high order n, [18]. Usually—although not always, [19, 20]—one chooses a *coordinate cross-section* obtained by setting (or normalizing) $r = \dim G$ of the jet coordinates equal to suitable constants. Almost always, one chooses a *minimal order* cross-section, meaning that the normalized jet coordinates have as low an order as possible. For example, the cross-section for an ordinary planar group action used in Sect. 2 is minimal. From here on we implicitly assume that we have chosen a coordinate cross-section of minimal order, although the general moving frame constructions can be readily adapted to more exotic choices.

If the group acts transitively on M, a minimal order coordinate cross-section is contained in the jet space over a single point $\mathscr{K} \subset \mathrm{J}^n|_{z_0}$. One can interpret such a coordinate cross-section as normalizing particular Taylor coefficients of the submanifolds passing through the base point z_0 — which is almost always taken to be at the origin. Once the moving frame map is specified, the *normal form* for a submanifold $S \subset M$ at a point $z \in S$ is obtained by applying the moving frame map corresponding to the submanifold's n-jet at the point in question, $z^{(n)} = \mathrm{j}_n S|_z$, so that $S_0 = \rho(z^{(n)}) \cdot S$ is a submanifold passing through $z_0 \in S_0$ and whose jet belongs to the cross-section, i.e., whose Taylor coefficients corresponding to the normalized cross-section jet coordinates have been normalized to the specified values. The remaining jet coordinates (Taylor coefficients), when expressed in terms of the originating submanifold jets $z^{(k)}$, provide a complete system of differential invariants, known as the *normalized differential invariants*.

With the moving frame in hand, we define the *invariantization* of a differential function $F(z^{(n)})$ to be the unique differential invariant $J(z^{(n)}) = \iota\big[\, F(z^{(n)}) \,\big]$ that agrees with F on the cross-section: $F \mid \mathscr{K} = J \mid \mathscr{K}$. In particular, invariantization of the jet coordinate functions leads to the normalized differential invariants:

$$H^i = \iota(x^i), \qquad I^\alpha_J = \iota(u^\alpha_J). \tag{45}$$

The r jet coordinates that are used to define the cross-section produce the constant, phantom differential invariants, and the remaining, non-phantom fundamental

[13] As before, the notation allows ρ to be only defined on an open subset of J^n.

normalized invariants provide a complete system of functionally independent differential invariants. The invariantization map has the explicit formula

$$\iota\big[\, F(\ldots x^i \ldots u^\alpha_J \ldots) \,\big] = F(\ldots H^i \ldots I^\alpha_J \ldots), \tag{46}$$

in which one replaces all jet coordinates by the corresponding normalized differential invariants. Moreover, invariantization clearly preserves differential invariants, $\iota(J) = J$, and hence any differential invariant can be expressed in terms of the normalized differential invariants via the *Replacement Rule*:

$$J(\ldots x^i \ldots u^\alpha_J \ldots) = J(\ldots H^i \ldots I^\alpha_J \ldots). \tag{47}$$

Furthermore, for p-dimensional submanifolds there are p invariant differential operators $\mathscr{D}_1, \ldots, \mathscr{D}_p$ that map differential invariants to differential invariants, and obtained by invariantizing[14] the corresponding total derivative operators

$$D_i = \frac{\partial}{\partial x^i} + \sum_{\alpha=1}^{q} \sum_J u^\alpha_{J,i} \frac{\partial}{\partial u^\alpha_J}, \qquad i = 1, \ldots, p, \tag{48}$$

where $u^\alpha_{J,i} = D_i(u^\alpha_J) = u^\alpha_{j_1 \ldots j_k i}$. The Basis Theorem, [1, 3], states that there exist a finite number of generating differential invariants $J_1, \ldots, J_l$ with the property that any other differential invariant can be written as a (not necessarily uniquely specified) function of the generating invariants and their successive invariant derivatives,

$$J_{k,I} = \mathscr{D}_{i_1} \ldots \mathscr{D}_{i_n} J_k, \qquad k = 1, \ldots, l, \qquad 1 \le i_\nu \le p, \qquad n \ge 0.$$

In particular, one can express all the normalized differential invariants in terms of them, and the explicit formulae can be found by iteratively applying the recurrence formulae, to be described next. It is known that, given a moving frame $\rho : \mathrm{J}^n \to G$ of order n, the non-constant normalized differential invariants of order $\le n+1$ form a generating set, although it typically contains redundancies and one can, by inspection of the recurrence relations and the commutation formulae (see Example 5 below) among the invariant differential operators, produce a smaller generating set. Determining the minimal number $l = l_{\min}$ of generating differential invariants is a very challenging problem, with surprises even in seemingly well-studied situations, [21, 22]. The case of curves is, however, known, where the answer (for ordinary group actions) is precisely $l = m - 1$; see [23] for intriguing Lie theoretic tools for determining their orders.

[14] More correctly, one invariantizes the basic horizontal one-forms, $\varpi^i = \iota(dx^i)$, producing a invariant horizontal coframe, and the invariant differential operators are the dual total differentiation operators.

To write out the recurrence formulae, let

$$\mathbf{v}_\sigma = \sum_{i=1}^{p} \xi_\sigma^i(x,u)\frac{\partial}{\partial x^i} + \sum_{\alpha=1}^{q} \varphi_\sigma^\alpha(x,u)\frac{\partial}{\partial u^\alpha}, \qquad \sigma = 1,\ldots,r, \tag{49}$$

be a basis for the infinitesimal generators of the action of G on M. The corresponding prolonged infinitesimal generators for the action on the jet spaces are given by the well-known *prolongation formula*

$$\operatorname{pr}\mathbf{v}_\sigma = \sum_{i=1}^{p} \xi_\sigma^i(x,u)\frac{\partial}{\partial x^i} + \sum_{\alpha=1}^{q}\sum_{\#J\geq 0} \varphi_{J,\sigma}^\alpha(x,u^{(n)})\frac{\partial}{\partial u_J^\alpha}, \tag{50}$$

whose coefficients are readily calculated:

$$\varphi_J^\alpha = D_J\left[\varphi^\alpha - \sum_{i=1}^{p}\xi^i u_i^\alpha\right] + \sum_{i=1}^{p}\xi^i u_{J,i}^\alpha, \tag{51}$$

where $D_J = D_{j_1}\cdot \ldots \cdot D_{j_k}$, with $J = (j_1,\ldots,j_k)$, $1 \leq j_\nu \leq p$, denotes the corresponding higher order total derivative.

The general *recurrence formula* for differential invariants[15] can be then formulated as follows. Let $F(z^{(n)})$ be any differential function. Then

$$\iota(D_i F) = \mathscr{D}_i\iota(F) - \sum_{\sigma=1}^{r} K_i^\sigma\,\iota\big[\operatorname{pr}\mathbf{v}_\sigma(F)\big], \qquad i = 1,\ldots,p, \tag{52}$$

where[16] K_i^σ are certain differential invariants known as the *Maurer–Cartan invariants*. (Our earlier equations (26), (27) are both special cases of (52), in which $F = u_k$ and x, respectively.) In particular, if we take F to be one of the cross-section coordinates, then its invariantization is a constant phantom invariant, and hence the first term on the hand side of (52) is zero. Thus, fixing $1 \leq i \leq p$, and then successively substituting the r cross-section coordinates into (52) produces a system of $r = \dim G$ linear equations which, according to [1], can be uniquely solved for the Maurer–Cartan invariants $K_i^1,\ldots,K_i^r$ as rational functions of the normalized differential invariants. Substituting these expressions, for all $i = 1,\ldots,p$, into (52), where now F is taken to be successive non-normalized jet coordinates, produces the full system of recurrence relations that completely specifies the structure of the rational, non-commutative differential invariant algebra and, in particular, leads to the desired formulae for the Taylor coefficients as invariant derivatives of the generating differential invariants.

[15] This is a special case of the more general recurrence formula for differential forms, [1, 24].

[16] Now we have made the group index σ on the Maurer–Cartan invariants a superscript.

Let us illustrate the procedure with two further examples.

Example 4 Consider the $r = 6$ – dimensional Euclidean group SE(3) acting by rigid motions on space curves $C \subset M = \mathbb{R}^3$. Here the submanifolds have dimension $p = 1$, and we use coordinates $z = (x, u, v)$ on M. As usual, we concentrate on curves given by the graphs of functions: $u = u(x),\ v = v(x)$, although all our results can be readily adapted to general parametrized curves $z(t) = (\,x(t), u(t), v(t)\,)^T$. Indeed, the recurrence formulae and consequent relations among differential invariants make no reference as to how the curve is parametrized. On the other hand, when writing out explicit formulas for the differential invariants, we use

$$z_t = \begin{pmatrix} x_t \\ u_t \\ v_t \end{pmatrix} = \begin{pmatrix} 1 \\ u_x \\ v_x \end{pmatrix}, \quad z_{tt} = \begin{pmatrix} x_{tt} \\ u_{tt} \\ v_{tt} \end{pmatrix} = \begin{pmatrix} 0 \\ u_{xx} \\ v_{xx} \end{pmatrix}, \quad z_{ttt} = \begin{pmatrix} x_{ttt} \\ u_{ttt} \\ v_{ttt} \end{pmatrix} = \begin{pmatrix} 0 \\ u_{xxx} \\ v_{xxx} \end{pmatrix}, \tag{53}$$

and so on, to denote the derivative vectors along the curve, where the second expression can be used in the special case of a graph, parametrized by $t = x$.

A basis for the infinitesimal generators is provided by the vector fields

$$\begin{aligned} &\mathbf{v}_1 = \partial_x, && \mathbf{v}_2 = \partial_u, && \mathbf{v}_3 = \partial_v, \\ &\mathbf{v}_4 = v\,\partial_u - u\,\partial_v, && \mathbf{v}_5 = -u\,\partial_x + x\,\partial_u, && \mathbf{v}_6 = -v\,\partial_x + x\,\partial_v. \end{aligned} \tag{54}$$

Applying the prolongation formula (50), (24) leads to the corresponding prolonged infinitesimal generators on the curve jet spaces, which are parametrized by

$$x,\, u,\, v,\, u_x,\, v_x,\, u_{xx},\, v_{xx},\, u_{xxx},\, \ldots\, .$$

To order 3, we find

$$\begin{aligned} &\operatorname{pr}\mathbf{v}_1 = \partial_x, \qquad \operatorname{pr}\mathbf{v}_2 = \partial_u, \qquad \operatorname{pr}\mathbf{v}_3 = \partial_v, \\ &\operatorname{pr}\mathbf{v}_4 = v\,\partial_u - u\,\partial_v + v_x\,\partial_{u_x} - u_x\,\partial_{v_x} + v_{xx}\,\partial_{u_{xx}} - u_{xx}\,\partial_{v_{xx}} + v_{xxx}\,\partial_{u_{xxx}} - u_{xxx}\,\partial_{v_{xxx}} + \cdots, \\ &\operatorname{pr}\mathbf{v}_5 = -u\,\partial_x + x\,\partial_u + (1 + u_x^2)\,\partial_{u_x} + u_x v_x\,\partial_{v_x} + 3\,u_x u_{xx}\,\partial_{u_{xx}} + (u_{xx} v_x + 2\,u_x v_{xx})\,\partial_{v_{xx}} \\ &\qquad + (4\,u_x u_{xxx} + 3\,u_{xx}^2)\,\partial_{u_{xxx}} + (u_{xxx} v_x + 3\,u_{xx} v_{xx} + 3\,u_x v_{xxx})\,\partial_{v_{xxx}} + \cdots, \\ &\operatorname{pr}\mathbf{v}_6 = -v\,\partial_x + x\,\partial_v + u_x v_x\,\partial_{u_x} + (1 + v_x^2)\,\partial_{v_x} + (2\,u_{xx} v_x + u_x v_{xx})\,\partial_{u_{xx}} + 3\,v_x v_{xx}\,\partial_{v_{xx}} \\ &\qquad + (3u_{xxx} v_x + 3\,u_{xx} v_{xx} + u_x v_{xxx})\,\partial_{u_{xxx}} + (4 v_x v_{xxx} + 3 v_{xx}^2)\,\partial_{v_{xxx}} + \cdots . \end{aligned} \tag{55}$$

The classical moving frame, [15], relies on the normalization equations

$$x = 0, \qquad u = 0, \qquad v = 0, \qquad u_x = 0, \qquad v_x = 0, \qquad v_{xx} = 0, \tag{56}$$

which serve to define a coordinate cross-section provided $u_{xx} \neq 0$. (Indeed, the classical moving frame is not defined at inflection points of the space curve.) This corresponds to translating and rotating the curve into the Euclidean normal form so that it goes through the origin, has tangent in the direction of the x-axis, and second order contact with the x, u plane. For this particular cross-section, the translational component of the left moving frame is the point on the curve, $z = (x, u, v) \in C$, while the columns of the rotational component $R = [\,\mathbf{t}, \mathbf{n}, \mathbf{b}\,] \in \mathrm{SO}(3)$ are the usual orthonormal tangent, normal, and binormal frame vectors based at z. However, keep in mind that these explicit identifications are not required to generate the recurrence formulae for the differential invariants.

We let

$$H = \iota(x), \qquad I_k = \iota(u_k), \qquad J_k = \iota(v_k), \tag{57}$$

be the normalized differential invariants resulting from invariantization, so that, in view of (56), the phantom invariants are

$$\begin{aligned} &H = \iota(x) = 0, && I_0 = \iota(u) = 0, && J_0 = \iota(v) = 0, \\ &I_1 = \iota(u_x) = 0, && J_1 = \iota(v_x) = 0, && J_2 = \iota(v_{xx}) = 0. \end{aligned} \tag{58}$$

One can further identify

$$I_2 = \iota(u_{xx}) = \kappa, \qquad J_3 = \iota(v_{xxx}) = \kappa\,\tau \tag{59}$$

with, respectively, the classical curvature invariant,[17] and the product of curvature and torsion. These two invariants generate the differential invariant algebra through invariant differentiation with respect to arc length, and the recurrence formulae allow one to express the normalized invariants I_k, J_k in terms of curvature, torsion, and their successive arc-length derivatives: $\kappa, \tau, \kappa_s, \tau_s, \ldots$.

We note the classical formulas

$$\begin{aligned} ds &= \|\, z_t \,\|\, dt = \sqrt{1 + u_x^2 + v_x^2}\, dx, \\ \kappa &= \frac{\|\, z_t \times z_{tt} \,\|}{\|\, z_t \,\|^3} = \frac{\sqrt{(u_x v_{xx} - u_{xx} v_x)^2 + u_{xx}^2 + v_{xx}^2}}{(1 + u_x^2 + v_x^2)^{3/2}}, \\ \tau &= \frac{z_t \times z_{tt} \cdot z_{ttt}}{\|\, z_t \times z_{tt} \,\|^2} = \frac{u_{xx} v_{xxx} - u_{xxx} v_{xx}}{(u_x v_{xx} - u_{xx} v_x)^2 + u_{xx}^2 + v_{xx}^2}, \end{aligned} \tag{60}$$

which can be obtained by fully implementing the moving frame construction, [25]. The first expression is valid for arbitrary parametrized curves, and the second is for graphs. However, we emphasize that these explicit formulas are *not required* for us to determine the recurrence formulas, and hence the Taylor coefficients of the Euclidean normal form of a space curve.

[17] As in the planar version, there is an ambiguous sign resulting from a 180° rotation, and one usually sets $\kappa = |\, I_2 \,|$ to ensure full invariance. To avoid minor technicalities, we shall ignore this extra complication here, and refer the reader to [16] for further details.

In this example, the recurrence formulae (52) have the form

$$\iota(D_x F) = \mathscr{D}\,\iota(F) - \sum_{\sigma=1}^{6} K_\sigma\, \iota(\mathrm{pr}\,\mathbf{v}_\sigma(F)), \tag{61}$$

for any differential function $F(x, u, v, u_x, v_x, u_{xx}, \ldots)$, where $K_1, \ldots, K_6$ are the Maurer–Cartan invariants. Taking F in (61) to be, in turn, each of the cross-section jet coordinates $x, u, v, u_x, v_x, v_{xx}$ that define the phantom invariants (58) leads, via (55), to the linear system

$$\begin{aligned} 1 &= \mathscr{D}H - K_1 = -K_1, & 0 &= I_1 = \mathscr{D}I_0 - K_2 = -K_2,\\ I_2 &= \mathscr{D}I_1 - K_5 = -K_5, & 0 &= J_1 = \mathscr{D}J_0 - K_3 = -K_3,\\ 0 = J_2 &= \mathscr{D}J_1 - K_6 = -K_6, & J_3 &= \mathscr{D}J_2 - I_2 K_4 = -I_2 K_4, \end{aligned}$$

which can be immediately solved for the Maurer–Cartan invariants:

$$K_1 = -1, \quad K_2 = 0, \quad K_3 = 0, \quad K_4 = -J_3/I_2 = -\tau, \quad K_5 = -I_2 = -\kappa, \quad K_6 = 0.$$

Substituting these expressions into (61) and letting F range over the other jet coordinates produces the non-phantom recurrence formulae

$$\begin{aligned} I_3 &= \mathscr{D}I_2, & &\\ I_4 &= \mathscr{D}I_3 + 3\,I_2^3 - J_3^2/I_2, & J_4 &= \mathscr{D}J_3 + I_3 J_3/I_2,\\ I_5 &= \mathscr{D}I_4 + 10\,I_2^2 I_3 - J_3 J_4/I_2, & J_5 &= \mathscr{D}J_4 + 6\,I_2^2 J_3 - J_3 I_4/I_2, \end{aligned} \tag{62}$$

and so on. Starting with (59), and successively substituting into (62), we find

$$\begin{aligned} I_2 &= \kappa, & &\\ I_3 &= \kappa_s, & J_3 &= \kappa\,\tau,\\ I_4 &= \kappa_{ss} + 3\kappa^3 - \kappa\,\tau^2, & J_4 &= \kappa\,\tau_s + 2\kappa_s\,\tau,\\ I_5 &= \kappa_{ssss} - 3\kappa\,\tau\,\tau_s - 3\kappa_s\,\tau^2 + 19\kappa^2\kappa_s, & J_5 &= \kappa\,\tau_{ss} + 3\kappa_s\,\tau_s + 3\kappa_{ss}\,\tau\\ & & &\quad - \kappa\,\tau^3 + 9\kappa^3\tau. \end{aligned} \tag{63}$$

This implies that the Euclidean normal form of a space curve has Taylor expansion

$$\begin{aligned} u_0(x) &= \tfrac{1}{2}\kappa\,x^2 + \tfrac{1}{6}\kappa_s\,x^3 + \tfrac{1}{24}(\kappa_{ss} + 3\kappa^3 - \kappa\,\tau^2)x^4 +\\ &\qquad + \tfrac{1}{120}(\kappa_{ssss} - 3\kappa\,\tau\,\tau_s - 3\kappa_s\,\tau^2 + 19\kappa^2\kappa_s)x^5 + \cdots,\\ v_0(x) &= \tfrac{1}{6}\kappa\,\tau\,x^3 + \tfrac{1}{24}(2\tau\kappa_s + \kappa\tau_s)x^4 +\\ &\qquad + \tfrac{1}{120}(\kappa\,\tau_{ss} + 3\kappa_s\,\tau_s + 3\kappa_{ss}\,\tau - \kappa\,\tau^3 + 9\kappa^3\tau)x^5 + \cdots. \end{aligned} \tag{64}$$

Observe that if $\tau \equiv 0$, so that the curve is planar, then the first equation in (64) reduces to the planar normal form (33).

Example 5 Finally, we treat the action of the Euclidean group SE(3) on two-dimensional surfaces $S \subset M = \mathbb{R}^3$. Now $p = 2$, and we use coordinates $z = (x, y, u)$ on M. As usual, we focus our attention to surfaces given by the graphs of functions: $u = u(x, y)$. All our results can be readily adapted to general parametrized surfaces, and, as always, the final recurrence formulae make no reference to the underlying parametrization. We refer to [21, 24] for additional details. The surface jet space has coordinates

$$(x, y, u, u_x, u_y, u_{xx}, u_{xy}, u_{yy}, u_{xxx}, u_{xxy}, u_{xyy}, u_{yyy}, \ldots),$$

and, in general, we use u_{jk} to denote the jet coordinate corresponding to the partial derivative $\partial^{j+k} u / \partial x^j \partial y^k$.

The classical moving frame construction, [15], relies on the coordinate cross-section

$$x = y = u = u_x = u_y = u_{xy} = 0. \tag{65}$$

The corresponding phantom invariants are

$$\begin{aligned} \iota(x) &= 0, & \iota(y) &= 0, & I_{00} &= \iota(u) = 0, \\ I_{10} = \iota(u_x) &= 0, & I_{01} = \iota(u_y) &= 0, & I_{11} &= \iota(u_{xy}) = 0, \end{aligned} \tag{66}$$

where, in general, we denote the normalized differential invariants by

$$I_{jk} = \iota(u_{jk}), \qquad j, k \geq 0.$$

The fundamental differential invariants of lowest order are the *principal curvatures*

$$\kappa_1 = I_{20} = \iota(u_{xx}), \qquad \kappa_2 = I_{02} = \iota(u_{yy}), \tag{67}$$

and it can be shown — through inspection of the recurrence formulae — that they generate the algebra of differential invariants via invariant differentiation. Surprisingly, as explained below, they do *not* form a minimal generating set.

The selected cross-section (65) corresponds to translating and rotating the surface so that it acquires the Euclidean normal form by passing through the origin, having horizontal tangent plane, and so that the directions of principal curvature line up with the coordinate axes. This requires that the point $z \in S$ be *non-umbilic*, meaning that the two principal curvatures are unequal, $\kappa_1 \neq \kappa_2$, which is the standard non-degeneracy condition required for the existence of a well-defined Euclidean moving frame, [15]. (At a non-degenerate umbilic, one could, in principle, employ a higher order moving frame.) The *mean* and *Gaussian curvature* invariants

$$H = \tfrac{1}{2}(\kappa_1 + \kappa_2), \qquad K = \kappa_1 \kappa_2, \tag{68}$$

are often used as convenient alternatives to the principal curvature invariants, since they eliminate some of the residual discrete ambiguities in the moving frame. The resulting left moving frame consists of the point on the curve defining the translation component $a = z \in \mathbb{R}^3$, while the columns of the rotation matrix $R = [\,\mathbf{t}_1, \mathbf{t}_2, \mathbf{n}\,] \in \mathrm{SO}(3)$ contain the unit tangent vectors $\mathbf{t}_1, \mathbf{t}_2$ forming the *Darboux frame* on the surface, [15], along with the unit normal $\mathbf{n}$.

Higher order differential invariants are obtained by differentiation with respect to the diagonalizing dual *Darboux coframe* $\varpi^1 = \iota(dx)$, $\varpi^2 = \iota(dy)$. We let $\mathscr{D}_1, \mathscr{D}_2$ denote the dual invariant differential operators, which differentiate in the principal curvature directions, and defined so that the differential of any differential function F can be written in invariant form

$$dF = (\mathscr{D}_1 F)\,\varpi^1 + (\mathscr{D}_2 F)\,\varpi^2. \tag{69}$$

The invariant differential operators do not commute, but, rather satisfy the *commutation relation*

$$\big[\,\mathscr{D}_1, \mathscr{D}_2\,\big] = \mathscr{D}_1\,\mathscr{D}_2 - \mathscr{D}_2\,\mathscr{D}_1 = Y_2\,\mathscr{D}_1 - Y_1\,\mathscr{D}_2, \tag{70}$$

where

$$Y_1 = \frac{\kappa_{2,1}}{\kappa_1 - \kappa_2}, \qquad Y_2 = \frac{\kappa_{1,2}}{\kappa_2 - \kappa_1}, \tag{71}$$

are known as the *commutator invariants*, whose expressions can also be established using the full moving frame calculus, [24]. Note that the denominator in (71) vanishes at umbilic points on the surface, where the principal curvatures coincide $\kappa_1 = \kappa_2$, and the moving frame is not valid.

Setting F to be, successively, x, y, u_{jk} in the general formulae (52) produces the recurrence relations

$$\begin{aligned}
1 &= -\sum_{\sigma=1}^{6} K_1^\sigma\,\iota(\xi_\sigma), \quad 0 = -\sum_{\sigma=1}^{6} K_1^\sigma\,\iota(\eta_\sigma), \quad I_{j+1,k} = \mathscr{D}_1 I_{jk} - \sum_{\sigma=1}^{6} K_1^\sigma\,\iota(\varphi_\sigma^{jk}),\\
0 &= -\sum_{\sigma=1}^{6} K_2^\sigma\,\iota(\xi_\sigma), \quad 1 = -\sum_{\sigma=1}^{6} K_2^\sigma\,\iota(\eta_\sigma), \quad I_{j,k+1} = \mathscr{D}_2 I_{jk} - \sum_{\sigma=1}^{6} K_2^\sigma\,\iota(\varphi_\sigma^{jk}),
\end{aligned} \tag{72}$$

for $j, k \geq 0$, where K_1^σ, K_2^σ are the Maurer–Cartan invariants, while $\xi_\sigma, \eta_\sigma, \varphi_\sigma^{jk}$ are, respectively, the coefficients of $\partial_x, \partial_y, \partial_{u_{jk}}$ in the prolonged infinitesimal generators, which are calculated via (51):

$$\begin{aligned}
\mathrm{pr}\,\mathbf{v}_1 &= \partial_x, \qquad \mathrm{pr}\,\mathbf{v}_2 = \partial_y, \qquad \mathrm{pr}\,\mathbf{v}_3 = \partial_u,\\
\mathrm{pr}\,\mathbf{v}_4 &= -\,y\,\partial_x + x\,\partial_y - u_y\partial_{u_x} + u_x\partial_{u_y}\\
&\qquad - 2u_{xy}\partial_{u_{xx}} + (u_{xx} - u_{yy})\,\partial_{u_{xy}} - 2u_{xy}\partial_{u_{yy}} + \cdots,\\
\mathrm{pr}\,\mathbf{v}_5 &= -\,u\,\partial_x + x\,\partial_u + (1 + u_x^2)\,\partial_{u_x} + u_x\,u_y\partial_{u_y}\\
&\qquad + 3u_x\,u_{xx}\partial_{u_{xx}} + (u_y\,u_{xx} + 2u_x\,u_{xy})\,\partial_{u_{xy}} + (2u_y\,u_{xy} + u_x\,u_{yy})\,\partial_{u_{yy}} + \cdots,
\end{aligned} \tag{73}$$

$$\text{pr}\,\mathbf{v}_6 = -\,u\,\partial_y + y\,\partial_u + u_x\,u_y\partial_{u_x} + (1+u_y^2)\,\partial_{u_y} + (u_y\,u_{xx} + 2u_x\,u_{xy})\,\partial_{u_{xx}} + (2u_y\,u_{xy} + u_x\,u_{yy})\,\partial_{u_{xy}} + 3u_y\,u_{yy}\partial_{u_{yy}} + \cdots .$$

Substituting (73) into the recurrence formulae (72) corresponding to the phantom invariants (58), and solving the resulting linear systems produces the formulae for the Maurer–Cartan invariants

$$\begin{aligned} &K_1^1 = -1, \quad K_1^2 = 0, \quad K_1^3 = 0, \quad K_1^4 = -Y_1, \quad K_1^5 = -\kappa_1, \quad K_1^6 = 0,\\ &K_2^1 = 0, \quad K_2^2 = -1, \quad K_2^3 = 0, \quad K_2^4 = -Y_2, \quad K_2^5 = 0, \quad K_2^6 = -\kappa_2. \end{aligned} \tag{74}$$

Substituting these expressions back into (72), we successively obtain the desired formulae for the higher order normalized differential invariants in terms of the principal curvatures, of which the third order ones are

$$I_{30} = \mathscr{D}_1\kappa_1 = \kappa_{1,1}, \quad I_{21} = \mathscr{D}_2\kappa_1 = \kappa_{1,2}, \quad I_{12} = \mathscr{D}_1\kappa_2 = \kappa_{2,1}, \quad I_{03} = \mathscr{D}_2\kappa_2 = \kappa_{2,2}, \tag{75}$$

while, taking these into account, the fourth order recurrence relations yield

$$\begin{aligned} I_{40} &= \kappa_{1,11} - \frac{3\kappa_{1,2}^2}{\kappa_1-\kappa_2} + 3\kappa_1^3,\\ I_{31} &= \kappa_{1,12} - \frac{3\kappa_{1,2}\kappa_{2,1}}{\kappa_1-\kappa_2} = \kappa_{1,21} + \frac{\kappa_{1,1}\kappa_{1,2} - 2\kappa_{1,2}\kappa_{2,1}}{\kappa_1-\kappa_2},\\ I_{22} &= \kappa_{1,22} + \frac{\kappa_{1,1}\kappa_{2,1} - 2\kappa_{2,1}^2}{\kappa_1-\kappa_2} + \kappa_1\kappa_2^2 = \kappa_{2,11} - \frac{\kappa_{1,2}\kappa_{2,2} - 2\kappa_{1,2}^2}{\kappa_1-\kappa_2} + \kappa_1^2\kappa_2,\\ I_{13} &= \kappa_{2,21} + \frac{3\kappa_{1,2}\kappa_{2,1}}{\kappa_1-\kappa_2} = \kappa_{2,12} - \frac{\kappa_{2,1}\kappa_{2,2} - 2\kappa_{1,2}\kappa_{2,1}}{\kappa_1-\kappa_2},\\ I_{04} &= \kappa_{2,22} + \frac{3\kappa_{2,1}^2}{\kappa_1-\kappa_2} + 3\kappa_2^3. \end{aligned} \tag{76}$$

There are two distinct formulae for I_{31}, I_{22}, I_{13} because they appear in both the first and second set of recurrence formulae in (72). The two expressions for I_{31} and I_{13} agree owing to the non-commutativity, (70), of $\mathscr{D}_1, \mathscr{D}_2$, while the two expressions for I_{22} imply the celebrated *Codazzi syzygy*

$$\kappa_{1,22} - \kappa_{2,11} + \frac{\kappa_{1,1}\,\kappa_{2,1} + \kappa_{1,2}\,\kappa_{2,2} - 2\kappa_{2,1}^2 - 2\kappa_{1,2}^2}{\kappa_1-\kappa_2} - \kappa_1\,\kappa_2\,(\kappa_1-\kappa_2) = 0, \tag{77}$$

which can be written compactly as

$$K = \kappa_1\kappa_2 = -\,(\mathscr{D}_1 + Y_1)Y_1 - (\mathscr{D}_2 + Y_2)Y_2. \tag{78}$$

The latter is the key identity employed by Guggenheimer, [15], for a short proof of Gauss' Theorema Egregium. We conclude that the Euclidean normal form of a surface $z = u(x, y)$ at a non-umbilic point is

$$u(x,y) = \tfrac{1}{2}\kappa_1 x^2 + \tfrac{1}{2}\kappa_2 y^2 + \tfrac{1}{6}\kappa_{1,1} x^3 + \tfrac{1}{2}\kappa_{1,2} x^2 y + \tfrac{1}{2}\kappa_{2,1} x y^2 + \tfrac{1}{6}\kappa_{2,2} y^3 + \tfrac{1}{24} I_{40} x^4 + \tfrac{1}{6} I_{31} x^3 y + \tfrac{1}{4} I_{22} x^2 y^2 + \tfrac{1}{6} I_{13} x y^3 + \tfrac{1}{24} I_{04} y^4 + \cdots, \tag{79}$$

where the fourth order coefficients appear in (76). Higher order terms can easily and automatically be determined using the recurrence formulae.

It is a classical result that the algebra of Euclidean differential invariants of a non-umbilic surface $S \subset \mathbb{R}^3$ is generated, through invariant differentiation, by the principal curvatures or, equivalently, the Gauss and mean curvatures; see [15] and, for a direct proof based on the moving frame recurrence relations, [24]. Surprisingly, as noted in [21], for suitably nondegenerate surfaces, the mean curvature by itself suffices to generate *all* the differential invariants. In particular, the Gauss curvature K can be written as an explicit universal rational combination of the invariant derivatives of the mean curvature H of order ≤ 4. Here we go slightly further by completely characterizing the nondegeneracy condition found in [21].

Definition 6 A surface $S \subset \mathbb{R}^3$ is *mean curvature degenerate* if, for any non-umbilic point $z_0 \in S$, there exist scalar functions $f_1(t)$, $f_2(t)$, such that

$$\mathscr{D}_1 H = f_1(H), \qquad \mathscr{D}_2 H = f_2(H), \tag{80}$$

at all points $z \in S$ in a suitable neighborhood of z_0.

Clearly any constant mean curvature surface—including any minimal surface—is mean curvature degenerate, with $f_1(t) \equiv f_2(t) \equiv 0$. Surfaces with non-constant mean curvature that admit a one-parameter group of Euclidean symmetries, i.e., non-cylindrical or non-spherical surfaces of rotation, non-planar surfaces of translation, or helicoid surfaces, obtained by, respectively, rotating, translating, or screwing a plane curve, are also mean curvature degenerate since, by the signature characterization of symmetry groups, [1], they have exactly one non-constant functionally independent differential invariant, namely their mean curvature H and hence any other differential invariant, including the invariant derivatives of H—as well as the Gauss curvature K—must be functionally dependent upon H. There also exist surfaces without continuous symmetries that are, nevertheless, mean curvature degenerate since it is entirely possible that (80) holds, but the Gauss curvature remains functionally independent of H. However, I do not know whether there is a good intrinsic geometric characterization of such surfaces.

Theorem 7 *If a surface is mean curvature nondegenerate then the algebra of differential invariants is generated entirely by the mean curvature and its successive invariant derivatives.*

Proof Following the arguments in [21], in view of the Codazzi formula (78), it suffices to write the commutator invariants Y_1, Y_2 in terms of the mean curvature. To this end, we note that the commutator identity (70) can be applied to any differential invariant. In particular,

$$\mathscr{D}_1\mathscr{D}_2 H - \mathscr{D}_2\mathscr{D}_1 H = Y_2\, \mathscr{D}_1 H - Y_1\, \mathscr{D}_2 H, \tag{81}$$

and, furthermore, for $j = 1$ or 2,

$$\mathscr{D}_1\mathscr{D}_2\mathscr{D}_j H - \mathscr{D}_2\mathscr{D}_1\mathscr{D}_j H = Y_2\, \mathscr{D}_1\mathscr{D}_j H - Y_1\, \mathscr{D}_2\mathscr{D}_j H. \tag{82}$$

Provided the nondegeneracy condition

$$(\mathscr{D}_1 H)(\mathscr{D}_2\mathscr{D}_j H) \neq (\mathscr{D}_2 H)(\mathscr{D}_1\mathscr{D}_j H), \quad \text{for } j = 1 \text{ or } 2, \tag{83}$$

holds, we can solve (81), (82) to write the commutator invariants Y_1, Y_2 as explicit rational functions of invariant derivatives of H. Plugging these expressions into the right hand side of the Codazzi identity (78) produces an explicit formula for the Gauss curvature as a rational function of the invariant derivatives, of order ≤ 4, of the mean curvature, which is valid for all surfaces satisfying the nondegeneracy condition (83).

Thus it remains to show that (83) is equivalent to mean curvature nondegeneracy of the surface. First, if (80) holds, then

$$\mathscr{D}_i\mathscr{D}_j H = \mathscr{D}_i f_j(H) = f_j'(H)\mathscr{D}_i H = f_j'(H) f_i(H), \qquad i, j = 1, 2.$$

This immediately implies

$$(\mathscr{D}_1 H)(\mathscr{D}_2\mathscr{D}_j H) = (\mathscr{D}_2 H)(\mathscr{D}_1\mathscr{D}_j H), \qquad j = 1, 2, \tag{84}$$

proving mean curvature degeneracy. Vice versa, in view of (69), the degeneracy condition (84) implies that, for each $j = 1, 2$, the differentials $dH, d(\mathscr{D}_j H)$ are linearly dependent everywhere on S, which, by a general theorem characterizing functional dependency, [6, Theorem 2.16], implies that, locally, $\mathscr{D}_j H$ can be written as a function of H, thus establishing the degeneracy condition (80). *Q.E.D.*

References

1. Fels, M., Olver, P.J.: Moving coframes. II. regularization and theoretical foundations. Acta Appl. Math. **55**, 127–208 (1999)
2. Olver, P.J., Pohjanpelto, J.: Moving frames for Lie pseudo-groups. Can. J. Math. **60**, 1336–1386 (2008)
3. Olver, P.J., Pohjanpelto, J.: Differential invariant algebras of Lie pseudo-groups. Adv. Math. **222**, 1746–1792 (2009)
4. Bruce, J.W., Giblin, P.J.: Curves and Singularities. Cambridge University Press, Cambridge (1984)
5. Cipolla, R., Giblin, P.: Visual Motion of Curves and Surfaces. Cambridge University Press, Cambridge (2000)
6. Olver, P.J.: Applications of Lie Groups to Differential Equations. Graduate texts in mathematics, 2nd edn., vol. 107. Springer, New York (1993)

7. Olver, P.J.: Equivalence, Invariants, and Symmetry. Cambridge University Press, Cambridge (1995)
8. Calabi, E., Olver, P.J., Shakiban, C., Tannenbaum, A., Haker, S.: Differential and numerically invariant signature curves applied to object recognition. Int. J. Comput. Vis. **26**, 107–135 (1998)
9. Grim, A., O'Connor, T., Olver, P.J., Shakiban, C., Slechta, R., Thompson, R.: Automatic reassembly of three–dimensional jigsaw puzzles. Int. J. Image Graph. **16**, 1650009 (2016)
10. Grim, A., Shakiban, C.: Applications of signatures in diagnosing breast cancer. Minnesota J. Undergrad. Math. **1**(1), 001 (2015)
11. Hoff, D., Olver, P.J.: Automatic solution of jigsaw puzzles. J. Math. Imaging Vis. **49**, 234–250 (2014)
12. Marsland, S., McLachlan, R.I.: Möbius invariants of shapes and images. SIGMA: Symmetry Integr. Geom. Methods Appl. **12**, 080 (2016)
13. Shakiban, C., Lloyd, P.: Signature curves statistics of DNA supercoils. In: Mladenov, I.M., Hirschfeld, A.C. (eds.) Geometry, Integrability and Quantization, vol. 5, pp. 203–210. Softex, Bulgaria (2004)
14. Hubert, E., Kogan, I.A.: Rational invariants of a group action. Construction and rewriting. J. Symb. Comput. **42**, 203–217 (2007)
15. Guggenheimer, H.W.: Differential Geometry. McGraw-Hill, New York (1963)
16. Olver, P.J.: Joint invariant signatures. Found. Comput. Math. **1**, 3–67 (2001)
17. Olver, P.J.: Lectures on moving frames. In: Levi, D., Olver, P., Thomova, Z., Winternitz, P. (eds.) Symmetries and Integrability of Difference Equations. London Math. Soc. Lecture Note Series, vol. 381, pp. 207–246. Cambridge University Press, Cambridge (2011)
18. Olver, P.J.: Moving frames and singularities of prolonged group actions. Selecta Math. **6**, 41–77 (2000)
19. Kim, P.: Invariantization of numerical schemes using moving frames. BIT **47**, 525–546 (2007)
20. Mansfield, E.L.: Algorithms for symmetric differential systems. Found. Comput. Math. **1**, 335–383 (2001)
21. Olver, P.J.: Differential invariants of surfaces. Differ. Geom. Appl. **27**, 230–239 (2009)
22. Hubert, E., Olver, P.J.: Differential invariants of conformal and projective surfaces. SIGMA **3**, 097 (2007)
23. Green, M.L.: The moving frame, differential invariants and rigidity theorems for curves in homogeneous spaces. Duke Math. J. **45**, 735–779 (1978)
24. Kogan, I.A., Olver, P.J.: Invariant Euler-Lagrange equations and the invariant variational bicomplex. Acta Appl. Math. **76**, 137–193 (2003)
25. Gray, A., Abbena, E., Salamon, S.: Modern Differential Geometry of Curves and Surfaces with Mathematica, 3rd edn. Chapman & Hall/CRC, FL (2006)

Integrable Nonlocal Reductions

Metin Gürses and Aslı Pekcan

Abstract We present some nonlocal integrable systems by using the Ablowitz–Musslimani nonlocal reductions. We first present all possible nonlocal reductions of nonlinear Schrödinger (NLS) and modified Korteweg–de Vries (mKdV) systems. We give soliton solutions of these nonlocal equations by using the Hirota method. We extend the nonlocal NLS equation to nonlocal Fordy–Kulish equations by utilizing the nonlocal reduction to the Fordy–Kulish system on symmetric spaces. We also consider the super AKNS system and then show that Ablowitz–Musslimani nonlocal reduction can be extended to super integrable equations. We obtain new nonlocal equations namely nonlocal super NLS and nonlocal super mKdV equations.

Keywords Ablowitz–Musslimani type reductions · Nonlocal NLS and mKdV equations · Hirota bilinear method · Soliton solutions · Nonlocal Fordy–Kulish system · Nonlocal super integrable NLS and mKdV equations

1 Introduction

After the publications of the Ablowitz–Musslimani works [1–3] on nonlocal nonlinear Schrödinger (NLS) equation there is a huge interest in obtaining nonlocal reductions of systems of integrable equations [5–8, 11–14, 21, 23–25, 28–34]. In all these works the soliton solutions and their properties were investigated by using inverse scattering techniques, by Darboux transformations, and by the Hirota direct method.

M. Gürses (✉)
Faculty of Science, Department of Mathematics, Bilkent University,
06800 Ankara, Turkey
e-mail: gurses@fen.bilkent.edu.tr

A. Pekcan
Faculty of Science, Department of Mathematics, Hacettepe University,
06800 Ankara, Turkey
e-mail: aslipekcan@hacettepe.edu.tr

V. G. Kac et al. (eds.), *Symmetries, Differential Equations and Applications*,
Springer Proceedings in Mathematics & Statistics 266,
https://doi.org/10.1007/978-3-030-01376-9_2

Recently we extended the nonlocal NLS equations to nonlocal Fordy–Kulish equations by utilizing the nonlocal reduction to the Fordy–Kulish system on symmetric spaces [15]. In a previous work [19] we studied the coupled NLS system obtained from AKNS scheme. By using the Hirota bilinear method we first found soliton solutions of the coupled NLS system of equations then using the Ablowitz–Musslimani type reduction formulas we obtained the soliton solutions of the standard and time T-, space S-, and space-time ST- reversal symmetric nonlocal NLS equations. Similarly, in a recent work [20] we studied the nonlocal modified Korteweg–de Vries (mKdV) equations which are also obtained from AKNS scheme by Ablowitz–Musslimani type nonlocal reductions. For this purpose we start using the soliton solutions of the coupled mKdV system found by Hirota and Iwao [22]. Then by using these solutions and Ablowitz–Musslimani type reduction formulas we obtained solutions of standard and nonlocal mKdV and complex mKdV (cmKdV) equations including one-, two-, and three-soliton waves, complexitons, breather-type, and kink-type waves. We used two different types of approaches in finding the soliton solutions. We gave one-soliton solutions of both types and presented only first type of two- and three-soliton solutions (see [20]).

When the Lax pair, in $(1+1)$-dimensions, is given in a Lie algebra the resulting evolution equations are given as a coupled system

$$q_t^i = F^i(q^k, r^k, q_x^k, r_x^k, q_{xx}^k, r_{xx}^k, \dots), \tag{1}$$
$$r_t^i = G^i(q^k, r^k, q_x^k, r_x^k, q_{xx}^k, r_{xx}^k, \dots), \tag{2}$$

for all $i = 1, 2, \dots, N$ where F^i and G^i are functions of the dynamical variables $q^i(t,x)$, $r^i(t,x)$, and their partial derivatives with respect to x. Since we start with a Lax pair then the system (1)–(2) is an integrable system of nonlinear partial differential equations.

In the space of dynamical variables (q^i, r^i) there exist subspaces

$$r^i(t,x) = kq^i(t,x), \tag{3}$$

or

$$r^i(t,x) = k\bar{q}^i(t,x), \tag{4}$$

where k is a constant and a bar over a letter denotes complex conjugation, such that the systems of equations (1)–(2) reduce to one system for q^i's

$$q_t^i = \tilde{F}^i(q^k, q_x^k, q_{xx}^k, \dots) \tag{5}$$

provided that the second system (2) consistently reduces to the above system (5) of equations. Here $\tilde{F} = F|_{r=k\bar{q}}$. Recently a new reduction is introduced by Ablowitz and Musslimani [1–3]

$$r^i(t,x) = kq^i(\mu_1 t, \mu_2 x), \tag{6}$$

or

$$r^i(t,x) = k\bar{q}^i(\mu_1 t, \mu_2 x), \tag{7}$$

for $i = 1, 2, \ldots, N$. Here k is a constant and $\mu_1^2 = \mu_2^2 = 1$. When $(\mu_1, \mu_2) = \{(-1, 1), (1, -1), (-1, -1)\}$ the above constraints reduce the system (1) to nonlocal differential equations provided that the second system (2) consistently reduces to the first one. If the reduction is done in a consistent way the reduced system of equations is also integrable. This means that the reduced system admits a recursion operator and bi-hamiltonian structure and the reduced system has N-soliton solutions. The inverse scattering method (ISM) can also be applied. Ablowitz and Musslimani have first found the nonlocal NLS equation from the coupled AKNS equations and solved it by ISM [2].

In our studies of nonlocal NLS and nonlocal mKdV equations we introduced a general method to obtain soliton solutions of nonlocal integrable equation. This method consists of three main steps:

(i) Find a consistent reduction formula which reduces the integrable system of equations to integrable nonlocal equations.

(ii) Find soliton solutions of the system of equations by use of the Hirota direct method or by inverse scattering transform technique, or by use of Darboux Transformation.

(iii) Use the reduction formulas on the soliton solutions of the system of equations to obtain the soliton solutions of the reduced nonlocal equations. By this way one obtains many different relations among the soliton parameters of the system of equations.

In the following sections we mainly follow the above method in obtaining the soliton solutions of the nonlocal NLS and nonlocal mKdV equations.

2 AKNS System

When we begin with the Lax pair in $sl(2, R)$ algebra and assume them as a polynomial of the spectral parameter of degree less or equal to three then we obtain the following system of evolution equations [4]:

$$q_t = a_2 \left(-\frac{1}{2} q_{xx} + q^2 r\right) + \mathrm{i}a_3 \left(-\frac{1}{4} q_{xxx} + \frac{3}{2} qrq_x\right), \tag{8}$$

$$r_t = a_2 \left(\frac{1}{2} r_{xx} - q r^2\right) + \mathrm{i}a_3 \left(-\frac{1}{4} r_{xxx} + \frac{3}{2} qrr_x\right). \tag{9}$$

Here a_2 and a_3 are arbitrary constants.

Letting $a_2 = 1/a$ and $a_3 = 0$ we get the coupled NLS system,

$$aq_t = -\frac{1}{2} q_{xx} + q^2 r, \tag{10}$$

$$ar_t = \frac{1}{2} r_{xx} - q r^2, \tag{11}$$

where a is any constant. The corresponding recursion operator is

$$\mathcal{R} = \begin{pmatrix} q\, D_x^{-1} r - \frac{1}{2} D_x & q\, D_x^{-1} q \\ -r\, D_x^{-1} r & -r\, D_x^{-1} q + \frac{1}{2} D_x \end{pmatrix}. \tag{12}$$

One-soliton solution of the system (10)–(11) can be obtained by the Hirota method as

$$q(t,x) = \frac{e^{\theta_1}}{1 + Ae^{\theta_1+\theta_2}}, \qquad r(t,x) = \frac{e^{\theta_2}}{1 + Ae^{\theta_1+\theta_2}}, \tag{13}$$

where $\theta_i = k_i x + \omega_i t + \delta_i,\ i = 1,2$ with $\omega_1 = k_1^2/2a,\ \omega_2 = -k_2^2/2a,$ and $A = -1/(k_1+k_2)^2$. Here k_1, k_2, δ_1, and δ_2 are arbitrary complex numbers.

3 Standard and Nonlocal NLS Equations

Standard reduction of NLS equation is $r(t,x) = k\bar{q}(t,x)$ where k is a real constant. The second equation (11) is consistent if $\bar{a} = -a$. Then the NLS system reduces to

$$aq_t = -\frac{1}{2} q_{xx} + k\, q^2\, \bar{q}. \tag{14}$$

Recursion operator of the NLS equation is

$$\mathcal{R} = \begin{pmatrix} kq\, D_x^{-1} \bar{q} - \frac{1}{2} D_x & q\, D_x^{-1} q \\ -k^2 \bar{q}\, D_x^{-1} \bar{q} & -\bar{q}\, D_x^{-1} q + \frac{1}{2} D_x \end{pmatrix}. \tag{15}$$

There are two types of approaches to find solutions of the standard and nonlocal NLS equations. In Type 1, one-soliton solution is obtained by letting $k_2 = \bar{k}_1$ and $e^{\delta_2} = ke^{\bar{\delta}_1}$ in (13) as

$$q(t,x) = \frac{e^{\theta_1}}{1 + A\,k\, e^{\theta_1+\bar{\theta}_1}}. \tag{16}$$

In Type 2 we obtain a different solution under the constraints,

$$(1)\ \bar{a} = -a,\ 2)\ k_1 = -\bar{k}_1,\ 3)\ k_2 = -\bar{k}_2,\ (4)\ Ake^{\delta_1+\bar{\delta}_1} = 1,\ (5)\ Ae^{\delta_2+\bar{\delta}_2} = k. \tag{17}$$

If we take $a = i\alpha$, $k_1 = i\beta$, $k_2 = i\gamma$, $e^{\delta_1} = a_1 + ib_1$, and $e^{\delta_2} = a_2 + ib_2$ for α, β, $\gamma, a_j, b_j \in \mathbb{R}$, $j = 1,2$ one-soliton solution of standard NLS equation becomes

$$q(t,x) = \frac{e^{i\beta x + \frac{i\beta^2}{2\alpha}t}(a_1 + ib_1)}{1 + \frac{1}{(\beta+\gamma)^2} e^{i(\beta+\gamma)x + i\frac{(\beta^2-\gamma^2)}{2\alpha}t}(a_1 + ib_1)(a_2 + ib_2)}, \quad \beta \neq -\gamma, \tag{18}$$

and therefore

$$|q(t,x)|^2 = \frac{a_1^2 + b_1^2}{4} \sec^2\left(\frac{\theta}{2}\right), \tag{19}$$

where

$$\theta = (\beta + \gamma)x + \frac{1}{2\alpha}(\beta^2 - \gamma^2)t + \omega_0$$

for $\omega_0 = \arccos((a_1 a_2 - b_1 b_2)/(\beta+\gamma)^2)$ with $a_1^2 + b_1^2 = (\beta+\gamma)^2/k$ and $a_2^2 + b_2^2 = k(\beta+\gamma)^2$. This solution is singular for any choice of the parameters.

Let now $r(t,x) = k\bar{q}(\mu_1 t, \mu_2 x)$ where $\mu_1^2 = \mu_2^2 = 1$ and k is a real constant. This is an integrable reduction, meaning that the new equation we obtain

$$aq_t(t,x) = -\frac{1}{2} q_{xx}(t,x) + k\, q^2(t,x)\, \bar{q}(\mu_1 t, \mu_2 x), \tag{20}$$

is integrable and the second equation (11) is consistent with the first one (10) provided that $\bar{a} = -\mu_1 a$. The recursion operator of this equation is

$$\mathscr{R} = \begin{pmatrix} k\, q(t,x)\, D_x^{-1}\, \bar{q}(\mu_1 t, \mu_2 x) - \frac{1}{2} D_x & q(t,x)\, D_x^{-1}\, q(t,x) \\ -k^2\, q(\mu_1 t, \mu_2 x)\, D_x^{-1}\, \bar{q}(\mu_1 t, \mu_2 x) & -k\, \bar{q}(\mu_1 t, \mu_2 x)\, D_x^{-1}\, q(t,x) + \frac{1}{2} D_x \end{pmatrix}, \tag{21}$$

and one-soliton solution is obtained by letting $k_2 = \mu_2 \bar{k}_1$ and $e^{\delta_2} = k e^{\bar{\delta}_1}$ in (13) as

$$q(t,x) = \frac{e^{\theta_1(t,x)}}{1 + A\,k\, e^{\theta_1(t,x) + \bar{\theta}_1(\mu_1 t, \mu_2 x)}}, \tag{22}$$

in Type 1 approach.

In Type 2, under the constraints

$$(1)\ \bar{a} = -\mu_1 a,\ (2)\ k_1 = -\bar{k}_1 \mu_2,\ (3)\ k_2 = -\bar{k}_2 \mu_2,\ (4)\ Ake^{\delta_1 + \bar{\delta}_1} = 1,\ (5)\ Ae^{\delta_2 + \bar{\delta}_2} = k, \tag{23}$$

we obtain a different one-soliton solution.

Nonlocal reductions of NLS system correspond to $(\mu_1, \mu_2) = \{(-1, 1), (1, -1), (-1, -1)\}$. Hence we have three different reductions of the NLS system (10)–(11).

(1) T-Symmetric Nonlocal NLS Equations: Let $r(t,x) = k\bar{q}(-t,x)$. This is an integrable equation

$$aq_t(t,x) = -\frac{1}{2} q_{xx}(t,x) + k\, q^2(t,x)\, \bar{q}(-t,x), \tag{24}$$

provided that $\bar{a} = a$. The recursion operator of this equation is

$$\mathscr{R} = \begin{pmatrix} k\, q(t,x)\, D_x^{-1}\, \bar{q}(-t,x) - \frac{1}{2}\, D_x & q(t,x)\, D_x^{-1}\, q(t,x) \\ -k^2\, \bar{q}(-t,x)\, D_x^{-1}\, \bar{q}(-t,x) & -k\, \bar{q}(-t,x)\, D_x^{-1}\, q(t,x) + \frac{1}{2}\, D_x \end{pmatrix}, \tag{25}$$

and one-soliton solution is obtained by letting $k_2 = \bar{k}_1$ where $k_1 = \alpha + i\beta, \alpha, \beta \in \mathbb{R}$, and $e^{\delta_2} = k e^{\bar{\delta}_1}$ in (13) as

$$q(t,x) = \frac{e^{(\alpha+i\beta)x + \frac{(\alpha+i\beta)^2}{2a}t + \delta_1}}{1 - k\frac{e^{2\alpha x + \frac{2i\alpha\beta}{a}t + \delta_1 + \bar{\delta}_1}}{4\alpha^2}}, \tag{26}$$

for $\alpha \neq 0$ in Type 1. To have a real-valued solution we consider $q(t,x)\bar{q}(t,x) = |q(t,x)|^2$. Here we have

$$|q(t,x)|^2 = \frac{16\alpha^4 e^{2\alpha x + \frac{\alpha^2-\beta^2}{a}t + \delta_1 + \bar{\delta}_1}}{(k e^{2\alpha x + \delta_1 + \bar{\delta}_1} - 4\alpha^2 \cos(\frac{2\alpha\beta}{a}t))^2 + 16\alpha^4 \sin^2(\frac{2\alpha\beta}{a}t)}. \tag{27}$$

When $\beta \neq 0$ and

$$t = \frac{an\pi}{2\alpha\beta}, \quad k e^{2\alpha x + \delta_1 + \bar{\delta}_1} - 4\alpha^2\,(-1)^n = 0,$$

where n is an integer, both focusing (sign $(k) = -1$) and defocusing (sign $(k) = 1$) cases have singularities. When $\beta = 0$ the focusing case is non-singular but asymptotically growing in time.

In Type 2, if we take $k_1 = i\beta$, $k_2 = i\gamma$ for $\beta, \gamma \in \mathbb{R}$, $e^{\delta_1} = a_1 + ib_1$, and $e^{\delta_2} = a_2 + ib_2$ for $a_j, b_j \in \mathbb{R}$, $j = 1, 2$ then one-soliton solution becomes

$$q(t,x) = \frac{e^{i\beta x - \frac{\beta^2}{2a}t}(a_1 + ib_1)}{1 + \frac{1}{(\beta+\gamma)^2} e^{i(\beta+\gamma)x + \frac{(\gamma^2-\beta^2)}{2a}t}(a_1 + ib_1)(a_2 + ib_2)}, \quad \beta \neq -\gamma. \tag{28}$$

Hence the function $|q(t,x)|^2$ is

$$|q(t,x)|^2 = \frac{e^{\frac{(\gamma^2+\beta^2)}{2a}t}(a_1^2 + b_1^2)}{2[\cosh(\frac{(\gamma^2-\beta^2)}{2a}t) + \cos\theta]}, \tag{29}$$

where

$$\theta = (\beta + \gamma)x + \omega_0$$

for $\omega_0 = \arccos((a_1a_2 - b_1b_2)/(\beta+\gamma)^2)$ with $a_1^2 + b_1^2 = (\beta+\gamma)^2/k$ and $a_2^2 + b_2^2 = k(\beta+\gamma)^2$. Clearly, the solution is singular at $t = 0$ and $\theta = (2n+1)\pi$, n integer and non-singular for $t \neq 0$.

(2) S-Symmetric Nonlocal NLS Equations: Let $r(t,x) = k\,\bar{q}(t,-x)$. This is an integrable equation

$$aq_t(t,x) = -\frac{1}{2}\,q_{xx}(t,x) + k\,q^2(t,x)\,\bar{q}(t,-x), \tag{30}$$

provided that $\bar{a} = -a$. The recursion operator of this equation is

$$\mathscr{R} = \begin{pmatrix} k\,q(t,x)\,D_x^{-1}\,\bar{q}(t,-x) - \frac{1}{2}\,D_x & q(t,x)\,D_x^{-1}\,q(t,x) \\ -k^2\,\bar{q}(t,-x)\,D_x^{-1}\,\bar{q}(t,-x) & -k\,\bar{q}(t,-x)\,D_x^{-1}\,q(t,x) + \frac{1}{2}\,D_x \end{pmatrix}. \tag{31}$$

In Type 1, one-soliton solution is obtained by letting $k_2 = -\bar{k}_1$ where $k_1 = \alpha + \mathrm{i}\beta$, $\alpha, \beta \in \mathbb{R}$, $a = \mathrm{i}y$, $y \in \mathbb{R}$, and $\mathrm{e}^{\delta_2} = k\mathrm{e}^{\bar{\delta}_1}$ in (13) as

$$q(t,x) = \frac{\mathrm{e}^{(\alpha+\mathrm{i}\beta)x + \frac{(\alpha+\mathrm{i}\beta)^2}{2iy}t + \delta_1}}{1 + k\dfrac{\mathrm{e}^{2\mathrm{i}\beta x + \frac{2\alpha\beta}{y}t + \delta_1 + \bar{\delta}_1}}{4\beta^2}}, \tag{32}$$

where $\beta \neq 0$. Hence the function $|q(t,x)|^2$ is

$$|q(t,x)|^2 = \frac{16\beta^4\mathrm{e}^{2\alpha x + \frac{2\alpha\beta}{y}t + \delta_1 + \bar{\delta}_1}}{(k\mathrm{e}^{\frac{2\alpha\beta}{y}t + \delta_1 + \bar{\delta}_1} + 4\beta^2\cos(2\beta x))^2 + 16\beta^4\sin^2(2\beta x)}. \tag{33}$$

If $\alpha \neq 0$ the above function is singular at

$$x = \frac{n\pi}{2\beta}, \quad k\mathrm{e}^{\frac{2\alpha\beta}{y}t + \delta_1 + \bar{\delta}_1} + 4\beta^2\,(-1)^n = 0,$$

where n is an integer, both for focusing and defocusing cases. If $\alpha = 0$, the function (33) becomes

$$|q(t,x)|^2 = \frac{2\beta^2}{k[B + \cos(2\beta x)]}, \tag{34}$$

for $B = (\rho^2 + 16\beta^4)/(8\rho\beta^2)$ where $\rho = k\mathrm{e}^{\delta_1 + \bar{\delta}_1}$. Obviously, the solution (34) is non-singular if $B > 1$ or $B < -1$.

Example 1 For the set of parameters
$(k_1, k_2, \mathrm{e}^{\delta_1}, \mathrm{e}^{\delta_2}, k, a) = (\mathrm{i}, \mathrm{i}, \mathrm{i}, -\mathrm{i}, 1, \mathrm{i}/2)$, we get the solution

$$|q(t,x)|^2 = \frac{16}{(17 + 8\cos(2x))}.$$

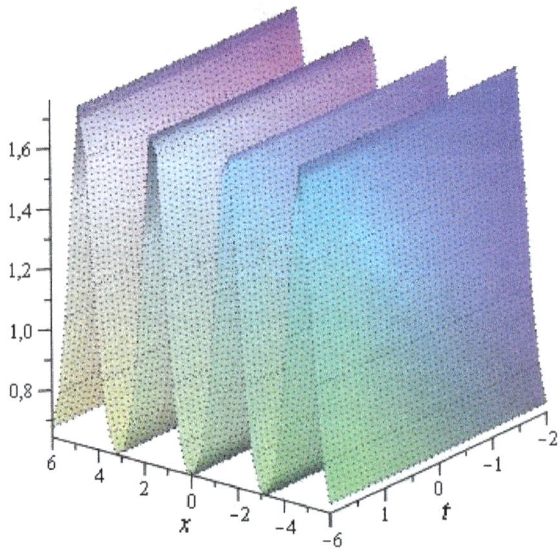

Fig. 1 A periodic solution corresponding to (34)

This solution represents a periodic solution. Its graph is given in Fig. 1.

For Type 2 if we let $a = \mathrm{i}\alpha, \alpha \in \mathbb{R}, \mathrm{e}^{\delta_1} = a_1 + \mathrm{i}b_1$, and $\mathrm{e}^{\delta_2} = a_2 + \mathrm{i}b_2$ for $a_j, b_j \in \mathbb{R}$, $j = 1, 2$ then one-soliton solution becomes

$$q(t,x) = \frac{\mathrm{e}^{k_1 x + \mathrm{i}\frac{k_1^2}{2\alpha}t}(a_1 + \mathrm{i}b_1)}{1 - \frac{1}{(k_1+k_2)^2}\mathrm{e}^{(k_1+k_2)x - \mathrm{i}\frac{(k_1^2-k_2^2)}{2\alpha}t}(a_1 + \mathrm{i}b_1)(a_2 + \mathrm{i}b_2)}, \quad k_1 \neq -k_2. \tag{35}$$

Therefore the function $|q(t,x)|^2$ is

$$|q(t,x)|^2 = \frac{\mathrm{e}^{(k_1-k_2)x}(a_1^2 + b_1^2)}{2[\cosh((k_1+k_2)x) - \cos\theta]}, \tag{36}$$

where

$$\theta = \frac{1}{2\alpha}(k_1^2 - k_2^2)t - \omega_0$$

for $\omega_0 = \arccos((a_1 a_2 - b_1 b_2)/(k_1 + k_2)^2)$ with $a_1^2 + b_1^2 = -(k_1 + k_2)^2/k$ and $a_2^2 + b_2^2 = -k(k_1 + k_2)^2$. Here $k_1, k_2 \in \mathbb{R}$. The solution is singular at $x = 0$ and $\theta = 2n\pi$ for n integer, and non-singular for $x \neq 0$.

(3) ST-Symmetric Nonlocal NLS Equations: Let $r(t,x) = k\,\bar{q}(-t,-x)$. This is an integrable equation

$$a\,q_t(t,x) = -\frac{1}{2}\,q_{xx}(t,x) + k\,q^2(t,x)\,\bar{q}(-t,-x), \tag{37}$$

provided that $\bar{a} = -a$. The recursion operator of this equation is

$$\mathscr{R} = \begin{pmatrix} k\, q(t,x)\, D_x^{-1}\, \bar{q}(-t,-x) - \frac{1}{2}\, D_x & q(t,x)\, D_x^{-1}\, q(t,x) \\ -k^2\, \bar{q}(-t,-x)\, D_x^{-1}\, \bar{q}(-t,-x) & -k\, \bar{q}(-t,-x)\, D_x^{-1}\, q(t,x) + \frac{1}{2}\, D_x \end{pmatrix}, \tag{38}$$

and one-soliton solution is obtained by letting $k_2 = -\bar{k}_1$ where $k_1 = \alpha + \mathrm{i}\beta, \alpha, \beta \in \mathbb{R}$ and $\mathrm{e}^{\delta_2} = k\mathrm{e}^{\bar{\delta}_1}$ in (13) as

$$q(t,x) = \frac{\mathrm{e}^{(\alpha+\mathrm{i}\beta)x + \frac{(\alpha+\mathrm{i}\beta)^2}{2a}t + \delta_1}}{1 + k\frac{\mathrm{e}^{2\mathrm{i}\beta x + \frac{2\mathrm{i}\alpha\beta}{a}t + \delta_1 + \bar{\delta}_1}}{4\beta^2}}, \tag{39}$$

where $\beta \neq 0$ in Type 1. Therefore $|q(t,x)|^2$ is

$$|q(t,x)|^2 = \frac{16\beta^4 \mathrm{e}^{2\alpha x + \frac{(\alpha^2-\beta^2)}{a}t + \delta_1 + \bar{\delta}_1}}{(k\mathrm{e}^{\delta_1+\bar{\delta}_1} + 4\beta^2\cos(2\beta x + \frac{2\alpha\beta}{a}t))^2 + 16\beta^4\sin^2(2\beta x + \frac{2\alpha\beta}{a}t)}. \tag{40}$$

This function is singular on the line $2\beta x + (2\alpha\beta t/a) = n\pi$ where n is an integer, if the condition $k\mathrm{e}^{\delta_1+\bar{\delta}_1} + 4\beta^2\,(-1)^n = 0$ is satisfied by the parameters of the solution, otherwise it represents a non-singular wave solution for both focusing and defocusing cases. For $\alpha = 0$, $(a > 0)$, the solution represents a localized wave solution. In Type 2, if we take $\mathrm{e}^{\delta_1} = a_1 + \mathrm{i}b_1$ and $\mathrm{e}^{\delta_2} = a_2 + \mathrm{i}b_2$ for $a_j, b_j \in \mathbb{R}$, $j = 1, 2$ we have the one-soliton solution as

$$q(t,x) = \frac{\mathrm{e}^{k_1 x + \frac{k_1^2}{2a}t}(a_1 + \mathrm{i}b_1)}{1 - \frac{1}{(k_1+k_2)^2}\mathrm{e}^{(k_1+k_2)x + (\frac{(k_1^2-k_2^2)}{2a})t}(a_1+\mathrm{i}b_1)(a_2+\mathrm{i}b_2)}, \quad k_1 \neq -k_2. \tag{41}$$

The corresponding function $|q(t,x)|^2$ is

$$|q(t,x)|^2 = \frac{\mathrm{e}^{\phi}(a_1^2 + b_1^2)}{1 - 2\gamma\mathrm{e}^{\theta} + \mathrm{e}^{2\theta}}, \tag{42}$$

where

$$\phi = 2k_1 x + \frac{k_1^2}{a}t, \quad \theta = (k_1 + k_2)x + \frac{1}{2a}(k_1^2 - k_2^2)t,$$

$\gamma = (a_1 a_2 - b_1 b_2)/(k_1 + k_2)^2$, $a_1^2 + b_1^2 = -(k_1 + k_2)^2/k$, and $a_2^2 + b_2^2 = -k(k_1 + k_2)^2$. Here $k_1, k_2 \in \mathbb{R}$. The above function is singular when the function $f(\theta) = \mathrm{e}^{2\theta} - 2\gamma\mathrm{e}^{\theta} + 1$ vanishes. It becomes zero when $\mathrm{e}^{\theta} = \gamma \pm \sqrt{\gamma^2 - 1}$. Hence if $\gamma < 1$ the solution is non-singular.

4 Standard and Nonlocal MKdV Equations

Letting $a_2 = 0$ and $a_3 = \mathrm{i}/a$ we get the mKdV system

$$aq_t = \frac{1}{4} q_{xxx} - \frac{3}{2} qrq_x, \tag{43}$$

$$ar_t = \frac{1}{4} r_{xxx} - \frac{3}{2} qrr_x. \tag{44}$$

This system has the same recursion operator (12) as the NLS system. One-soliton solution of the above system is [22]

$$q(t,x) = \frac{\mathrm{e}^{\theta_1}}{1 + A\mathrm{e}^{\theta_1+\theta_2}}, \qquad r(t,x) = \frac{\mathrm{e}^{\theta_2}}{1 + A\mathrm{e}^{\theta_1+\theta_2}}, \tag{45}$$

with $\theta_i = k_i x - (k_i^3 t/4a) + \delta_i$, $i = 1, 2$, and $A = -1/(k_1 + k_2)^2$. Here k_1, k_2, δ_1, and δ_2 are arbitrary complex numbers. In mKdV case, there are also two types of approaches represented in [20] to find solutions of the standard mKdV and nonlocal mKdV (and cmKdV) equations.

1. MKdV Equations: Let $r(t,x) = kq(t,x)$ then mKdV system reduces to the integrable mKdV equation

$$aq_t = \frac{1}{4} q_{xxx} - \frac{3k}{2} q^2 q_x. \tag{46}$$

In Type 1 one-soliton solution is obtained by letting $k_1 = k_2 = \alpha + \mathrm{i}\beta$ and $\mathrm{e}^{\delta_2} = k\mathrm{e}^{\delta_1} = a_1 + \mathrm{i}b_1$ for $\alpha, \beta, a_1, b_1 \in \mathbb{R}$ in (45) as

$$q(t,x) = \frac{\mathrm{e}^{(\alpha+\mathrm{i}\beta)x - \frac{(\alpha^3-3\alpha\beta^2)+\mathrm{i}(3\alpha^2\beta-\beta^3)}{4a}t}(a_1 + \mathrm{i}b_1)}{1 - \frac{k}{4(\alpha^2+\beta^2)^2}\mathrm{e}^{2(\alpha+\mathrm{i}\beta)x - \frac{(\alpha^3-3\alpha\beta^2)+\mathrm{i}(3\alpha^2\beta-\beta^3)}{2a}t}(a_1+\mathrm{i}b_1)^2(\alpha-\mathrm{i}\beta)^2}. \tag{47}$$

Therefore we obtain the function

$$|q(t,x)|^2 = \frac{Y}{W}, \tag{48}$$

where

$$Y = \mathrm{e}^{2\alpha x - \frac{(\alpha^3-3\alpha\beta^2)}{2a}t}(a_1^2 + b_1^2),$$
$$W = 1 - \gamma_1 \cos\theta + \frac{\gamma_1^2}{4}\mathrm{e}^{\phi} = \frac{\gamma_1^2}{4}\Big[\frac{4}{\gamma_1^2}(1 - \gamma_1\cos\theta) + \mathrm{e}^{\phi}\Big], \tag{49}$$

where

$$\theta = 2\beta x - \frac{1}{2a}(3\alpha^2\beta - \beta^3)t + \omega_0, \quad \phi = 4\alpha x - \frac{1}{a}(\alpha^3 - 3\alpha\beta^2)t,$$

for

$$\omega_0 = \arccos(((a_1\alpha + b_1\beta)^2 - (a_1\beta - b_1\alpha)^2)/(a_1^2 + b_1^2)(\alpha^2 + \beta^2))$$

and $\gamma_1 = k(a_1^2 + b_1^2)/2(\alpha^2 + \beta^2)$. Hence we conclude that if $|\gamma_1| \leq 1$ the solution (48) is non-singular. Type 2 approach gives $k_1 = k_2 = 0$ yielding trivial solution.

2. CmKdV Equations: Let $r = k\bar{q}(t, x)$ then mKdV system reduces to the integrable cmKdV equation

$$aq_t = \frac{1}{4}q_{xxx} - \frac{3k}{2}q\,\bar{q}\,q_x, \tag{50}$$

where $\bar{a} = a$. One-soliton solution is obtained by letting $k_2 = \bar{k}_1 = \alpha - i\beta$ for $\alpha, \beta \in \mathbb{R}$ and $e^{\delta_2} = ke^{\bar{\delta}_1}$ in (45) in Type 1 as

$$q(t, x) = \frac{e^{(\alpha+i\beta)x - \frac{(\alpha^3-3\alpha\beta^2)+i(3\alpha^2\beta-\beta^3)}{4a}t+\delta_1}}{1 - \frac{k}{4\alpha^2}e^{2\alpha x + \frac{(3\alpha\beta^2-\alpha^3)}{2a}t+\delta_1+\bar{\delta}_1}}, \tag{51}$$

so the function $|q(t, x)|^2$ is

$$|q(t, x)|^2 = \frac{e^{2\alpha x - \frac{(\alpha^3-3\alpha\beta^2)}{2a}t+\delta_1+\bar{\delta}_1}}{(1 - \frac{k}{4\alpha^2}e^{2\alpha x + \frac{(3\alpha\beta^2-\alpha^3)}{2a}t+\delta_1+\bar{\delta}_1})^2}. \tag{52}$$

For $k < 0$, the solution (52) can be written as

$$|q(t, x)|^2 = -\frac{\alpha^2}{k}\operatorname{sech}^2\Big(\alpha x + \frac{(3\alpha\beta^2 - \alpha^3)}{4a}t + \frac{\delta_1 + \bar{\delta}_1}{2} + \delta\Big), \tag{53}$$

where $\delta = \ln(-k/4\alpha^2)/2$. The above solution is non-singular.

We obtain a different one-soliton solution in Type 2 under the constraints $k_1 = -\bar{k}_1, k_2 = -\bar{k}_2$, $Ake^{\delta_1+\bar{\delta}_1} = 1$, and $Ae^{\delta_2+\bar{\delta}_2} = k$ used in (45). If we let $k_1 = i\alpha$, $k_2 = i\beta$, $e^{\delta_1} = a_1 + ib_1$, and $e^{\delta_2} = a_2 + ib_2$ for $\alpha, \beta, a_j, b_j \in \mathbb{R}$, $j = 1, 2$, one-soliton solution becomes

$$q(t, x) = \frac{e^{i\alpha x + i\frac{\alpha^3}{4a}t}(a_1 + ib_1)}{1 + \frac{1}{(\alpha+\beta)^2}e^{i(\alpha+\beta)x + i\frac{(\alpha^3+\beta^3)}{4a}t}(a_1 + ib_1)(a_2 + ib_2)}, \tag{54}$$

hence the corresponding function $|q(t, x)|^2$ is

$$|q(t,x)|^2 = \frac{a_1^2+b_1^2}{4}\sec^2\left(\frac{\theta}{2}\right), \tag{55}$$

where

$$\theta = (\alpha+\beta)x + \frac{1}{4a}(\alpha^3+\beta^3)t + \omega_0,$$

for $\omega_0 = \arccos((a_1a_2 - b_1b_2)/(\alpha+\beta)^2)$ with $a_1^2+b_1^2 = (\alpha+\beta)^2/k$ and $a_2^2 + b_2^2 = k(\alpha+\beta)^2$. This is a singular solution for $\theta = (2n+1)\pi$, n is an integer.

There are also two different types of nonlocal reductions.

1. Nonlocal MKdV Equations: Let $r = kq(\mu_1 t, \mu_2 x)$ then mKdV system reduces to the integrable nonlocal mKdV equation

$$aq_t(t,x) = \frac{1}{4}q_{xxx}(t,x) - \frac{3k}{2}q(t,x)q(\mu_1 t,\mu_2 x)q_x(t,x), \tag{56}$$

provided that $\mu_1\mu_2 = 1$. There is only one possibility $(\mu_1,\mu_2) = (-1,-1)$. If we consider the Type 1 approach, we get $k_1 = -k_2$ which gives trivial solution $q(t,x) = 0$. In Type 2, one-soliton solution is obtained from (45) with the parameters satisfying the relations $Ake^{2\delta_1} = 1$ and $Ae^{2\delta_2} = k$ as

$$q(t,x) = \frac{i\sigma_1 e^{k_1 x - \frac{k_1^3}{4a}t}(k_1+k_2)}{\sqrt{k}(1+\sigma_1\sigma_2 e^{(k_1+k_2)x - \frac{(k_1^3+k_2^3)}{4a}t})}, \quad \sigma_j = \pm 1, \quad j = 1,2. \tag{57}$$

If we let $a \in \mathbb{R}$, $k_1 = \alpha_1 + i\beta_1$, and $k_2 = \alpha_2 + i\beta_2$ then we obtain the solution $|q(t,x)|^2$ corresponding to (57) as

$$|q(t,x)|^2 = \frac{e^{\theta}((\alpha_1+\alpha_2)^2 + (\beta_1+\beta_2)^2)}{2k[\cosh(\phi) + \sigma_1\sigma_2\cos(\varphi)]}, \tag{58}$$

where $\theta = (\alpha_1 - \alpha_2)x - ((\alpha_1^3 - 3\alpha_1\beta_1^2 - \alpha_2^3 + 3\alpha_2\beta_2^2)t/4a)$, $\phi = A_1x + B_1t$, and $\varphi = A_2x + B_2t$. Here

$$A_1 = \alpha_1 + \alpha_2, \quad B_1 = -\frac{1}{4a}(\alpha_1^3 - 3\alpha_1\beta_1^2 + \alpha_2^3 - 3\alpha_2\beta_2^2),$$
$$A_2 = \beta_1 + \beta_2, \quad B_2 = \frac{1}{4a}(\beta_1^3 - 3\alpha_1^2\beta_1 + \beta_2^3 - 3\alpha_2^2\beta_2).$$

There are cases where the solution (58) is nonsingular:

(a) If we have $k_1 = k_2$ for real k_1 and $\sigma_1\sigma_2 = 1$ then the solution (57) becomes

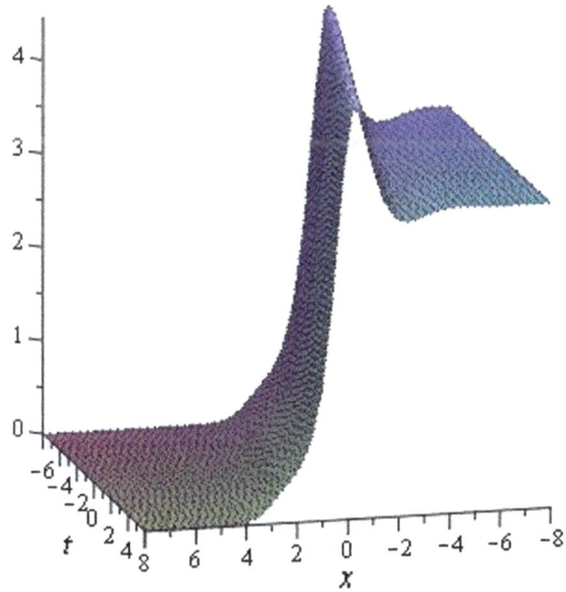

Fig. 2 A complexiton solution corresponding to (60)

$$q(t,x)=\frac{i\sigma_1 k_1}{\sqrt{k}}\operatorname{sech}(k_1 x-\frac{k_1^3}{4a}t). \tag{59}$$

(b) If $B_1A_2 = B_2A_1$ then the solution (58) becomes

$$|q(t,x)|^2=\frac{e^{\theta}((\alpha_1+\alpha_2)^2+(\beta_1+\beta_2)^2)}{2k[\cosh(\phi)+\sigma_1\sigma_2\cos(\frac{B_2}{B_1}\phi)]}. \tag{60}$$

Example 2 If we take $(k_1, k_2, \sigma_1, \sigma_2 k, a) = (\mathrm{i}, 1+(\mathrm{i}/2), 1, 1, -1, 1/4)$ then we have the solution

$$|q(t,x)|^2=\frac{13e^{-u}}{8[\cosh(u)+\cos(3u/2)]},$$

where $u = x - t/4$. This is a complexiton solution. The graph of this solution is given in Fig. 2.

2. Nonlocal CmKdV Equations: Let $r = k\bar{q}(\mu_1 t, \mu_2 x)$ then mKdV system reduces to the integrable nonlocal cmKdV equation

$$aq_t(t,x)=\frac{1}{4}q_{xxx}(t,x)-\frac{3k}{2}q(t,x)\bar{q}(\mu_1 t,\mu_2 x)q_x(t,x), \tag{61}$$

provided that $\bar{a} = \mu_1\mu_2 a$. One-soliton solution is obtained by letting $k_2 = \mu_2\bar{k}_1$ and $e^{\delta_2} = ke^{\bar{\delta}_1}$ in Type 1. In Type 2, a different one-soliton solution is obtained by letting $k_1 = -\bar{k}_1\mu_2$, $k_2 = -\bar{k}_2\mu_2$, $Ake^{\delta_1+\bar{\delta}_1} = 1$, and $Ae^{\delta_2+\bar{\delta}_2} = k$. In this case there are three possibilities $(\mu_1, \mu_2) = \{(-1, 1), (1, -1), (-1, -1)\}$. Hence we have three integrable nonlocal cmKdV equations:

2(i) T-Symmetric Nonlocal CmKdV Equations: Let $r = k\bar{q}(-t, x)$ then mKdV system reduces to the nonlocal cmKdV equation

$$aq_t(t, x) = -\frac{1}{4}q_{xxx}(t, x) + \frac{3}{2}k\bar{q}(-t, x)q(t, x)q_x(t, x), \quad \bar{a} = -a. \tag{62}$$

In Type 1 if we let $a = ib$, for nonzero $b \in \mathbb{R}$, $k_1 = \alpha + i\beta$ so $k_2 = \alpha - i\beta$ for $\alpha, \beta \in \mathbb{R}$, $\alpha \neq 0$ then one-soliton solution becomes

$$q(t, x) = \frac{e^{(\alpha+i\beta)x+\frac{i(\alpha^3-3\alpha\beta^2)-3\alpha^2\beta+\beta^3}{4b}t+\delta_1}}{1 - \frac{k}{4\alpha^2}e^{2\alpha x+i\frac{\alpha^3-3\alpha\beta^2}{2b}t+\delta_1+\bar{\delta}_1}}. \tag{63}$$

The corresponding function $|q(t, x)|^2$ is

$$|q(t, x)|^2 = \frac{e^{2\alpha x+\frac{(\beta^3-3\alpha^2\beta)}{2b}t+\delta_1+\bar{\delta}_1}}{\left[\frac{k}{4\alpha^2}e^{2\alpha x+\delta_1+\bar{\delta}_1} - \cos(\frac{(\alpha^3-3\alpha\beta^2)}{2b}t)\right]^2 + \sin^2(\frac{(\alpha^3-3\alpha\beta^2)}{2b}t)}. \tag{64}$$

when $\alpha^3 - 3\alpha\beta^2 \neq 0$ and

$$t = \frac{2nb\pi}{(\alpha^3 - 3\alpha\beta^2)}, \quad \frac{k}{4\alpha^2}e^{2\alpha x+\delta_1+\bar{\delta}_1} - (-1)^n = 0,$$

where n is an integer, for both focusing and defocusing cases, the solution is singular. When $\alpha^3 - 3\alpha\beta^2 = 0$ the solution for focusing case is non-singular. When $\alpha = 0$ the solution is exponentially growing for $\beta^3/b > 0$ and exponentially decaying for $\beta^3/b < 0$.

In Type 2 if we let $a = i\alpha$, $k_1 = i\beta$, $k_2 = i\gamma$ for $\alpha, \beta, \gamma \in \mathbb{R}$, and $e^{\delta_1} = a_1 + ib_1$, $e^{\delta_2} = a_2 + ib_2$ for $a_j, b_j \in \mathbb{R}$, $j = 1, 2$ then one-soliton solution becomes

$$q(t, x) = \frac{e^{i\beta x+\frac{\beta^3}{4\alpha}t}(a_1 + ib_1)}{1 + \frac{1}{(\beta+\gamma)^2}e^{i(\beta+\gamma)x+\frac{(\beta^3+\gamma^3)}{4\alpha}t}(a_1 + ib_1)(a_2 + ib_2)}. \tag{65}$$

Hence the function $|q(t, x)|^2$ is

$$|q(t, x)|^2 = \frac{e^{\frac{(\beta^3-\gamma^3)}{4\alpha}t}(a_1^2 + b_1^2)}{2\left[\cosh(\frac{(\beta^3+\gamma^3)}{4\alpha}t) + \cos\theta\right]}, \tag{66}$$

where

$$\theta = (\beta + \gamma)x + \omega_0$$

for $\omega_0 = \arccos((a_1a_2 - b_1b_2)/(\beta+\gamma)^2)$ with $a_1^2 + b_1^2 = (\beta+\gamma)^2/k$, $a_2^2 + b_2^2 = k(\beta+\gamma)^2$, and $\beta \neq -\gamma$. This solution is singular only at $t = 0$, $\theta = (2n+1)\pi$ for n integer.

2(ii) S-Symmetric Nonlocal CmKdV Equations: Let $r = k\bar{q}(t,-x)$ then mKdV system reduces to the nonlocal cmKdV equation

$$aq_t(t,x) = -\frac{1}{4}q_{xxx}(t,x) + \frac{3}{2}k\bar{q}(t,-x)q(t,x)q_x(t,x), \quad \bar{a} = -a. \tag{67}$$

If we consider Type 1 and let $a = ib$ for nonzero $b \in \mathbb{R}$, $k_1 = \alpha + i\beta$ and so $k_2 = -\alpha + i\beta$ for $\alpha, \beta \in \mathbb{R}$, $\beta \neq 0$ then one-soliton solution becomes

$$q(t,x) = \frac{e^{(\alpha+i\beta)x + \frac{i\alpha^3 - 3\alpha^2\beta - 3i\alpha\beta^2 + \beta^3}{4b}t + \delta_1}}{1 + \frac{k}{4\beta^2}e^{2i\beta x + i\frac{\alpha^3 - 3\alpha\beta^2}{2b}t + \delta_1 + \bar{\delta}_1}}, \tag{68}$$

and so $|q(t,x)|^2$ is

$$|q(t,x)|^2 = \frac{e^{2\alpha x + \frac{(\beta^3 - 3\alpha^2\beta)}{2b}t + \delta_1 + \bar{\delta}_1}}{\left[\frac{k}{4\beta^2}e^{\frac{(\beta^3 - 3\alpha^2\beta)}{2b}t + \delta_1 + \bar{\delta}_1} + \cos(2\beta x)\right]^2 + \sin^2(2\beta x)}. \tag{69}$$

For $x = n\pi/(2\beta)$ and $ke^{(\beta^3 - 3\alpha^2\beta)t/2b + \delta_1 + \bar{\delta}_1}/(4\beta^2) + (-1)^n = 0$, where n is an integer, the solution is unbounded but for $\beta^2 = 3\alpha^2$ and $ke^{\delta_1 + \bar{\delta}_1}/(4\beta^2) + (-1)^n \neq 0$ we have a periodical solution. For $\alpha = 0$, the solution (69) becomes

$$|q(t,x)|^2 = \frac{e^{\delta_1 + \bar{\delta}_1}}{\gamma[\sigma_k \cosh(\frac{\beta^3}{2b}t + \ln(\frac{|\gamma|}{2})) + \cos(2\beta x)]}, \tag{70}$$

where $\gamma = ke^{\delta_1 + \bar{\delta}_1}/(2\beta^2)$, $\sigma_k = 1$ if $k > 0$, and $\sigma_k = -1$ if $k < 0$. This solution is non-singular for $|\gamma| > 2$, $\beta^3/b > 0$ and $|\gamma| < 2$, $\beta^3/b < 0$ for any $t \geq 0$.

For Type 2 if we let $a = i\alpha, \alpha \in \mathbb{R}$, $e^{\delta_1} = a_1 + ib_1$, and $e^{\delta_2} = a_2 + ib_2$ for $a_j, b_j \in \mathbb{R}$, $j = 1, 2$ we obtain the one-soliton solution as

$$q(t,x) = \frac{e^{k_1 x + i\frac{k_1^3}{4\alpha}t}(a_1 + ib_1)}{1 - \frac{1}{(k_1+k_2)^2}e^{(k_1+k_2)x + i\frac{(k_1^3 + k_2^3)}{4\alpha}t}(a_1 + ib_1)(a_2 + ib_2)}. \tag{71}$$

Therefore the function $|q(t,x)|^2$ is

$$|q(t,x)|^2 = \frac{\mathrm{e}^{(k_1-k_2)x}(a_1^2+b_1^2)}{2[\cosh((k_1+k_2)x)+\cos\theta]}, \tag{72}$$

where

$$\theta = \frac{1}{4}(k_1^3+k_2^3)t - \omega_0$$

for $\omega_0 = \arccos((b_1b_2 - a_1a_2)/(k_1+k_2)^2)$ with $a_1^2+b_1^2 = -\frac{(k_1+k_2)^2}{k}$ and $a_2^2+b_2^2 = -k(k_1+k_2)^2$, $k_1 \neq -k_2$. Here $k_1, k_2 \in \mathbb{R}$. This solution has singularity at $x = 0$, $\theta = (2n+1)\pi$ for n integer.

2(iii) ST-Symmetric Nonlocal CmKdV Equations: Let $r = k\bar{q}(-t,-x)$ then mKdV system reduces to the nonlocal cmKdV equation

$$aq_t(t,x) = -\frac{1}{4}q_{xxx}(t,x) + \frac{3}{2}k\bar{q}(-t,-x)q(t,x)q_x(t,x), \quad \bar{a} = a. \tag{73}$$

In Type 1 if we let $k_1 = \alpha + \mathrm{i}\beta$ and so $k_2 = -\alpha + \mathrm{i}\beta$ for $\alpha, \beta \in \mathbb{R}$, $\beta \neq 0$ the one-soliton solution $q(t,x)$ becomes

$$q(t,x) = \frac{\mathrm{e}^{(\alpha+\mathrm{i}\beta)x - \frac{\alpha^3+3\alpha^2\mathrm{i}\beta-3\alpha\beta^2-\mathrm{i}\beta^3}{4a}t+\delta_1}}{1+\frac{k}{4\beta^2}\mathrm{e}^{2\mathrm{i}\beta x - \mathrm{i}\frac{(6\alpha^2\beta-2\beta^3)}{4a}t+\delta_1+\bar{\delta}_1}}. \tag{74}$$

Then we obtain the function $|q(t,x)|^2$ as

$$|q(t,x)|^2 = \frac{\mathrm{e}^{\theta}}{\mu\left[(\frac{1}{\mu}+\frac{\mu}{4})+\cos\phi\right]}, \tag{75}$$

where

$$\theta = 2\alpha x + \frac{1}{2a}(3\alpha\beta^2-\alpha^3)t + \delta_1 + \bar{\delta}_1, \phi = 2\beta x + \frac{1}{2a}(\beta^3 - 3\alpha^2\beta)t,$$

and $\mu = k\mathrm{e}^{\delta_1+\bar{\delta}_1}/(2\beta^2)$. This solution is non-singular for all μ except $\mu = \pm 2$.

For Type 2, if we take $\mathrm{e}^{\delta_1} = a_1 + ib_1$ and $\mathrm{e}^{\delta_2} = a_2 + ib_2$ for $a_j, b_j \in \mathbb{R}$, $j = 1, 2$ we obtain the one-soliton solution as

$$q(t,x) = \frac{\mathrm{e}^{k_1x - \frac{k_1^3}{4a}t}(a_1+\mathrm{i}b_1)}{1 - \frac{1}{(k_1+k_2)^2}\mathrm{e}^{(k_1+k_2)x - \frac{(k_1^3+k_2^3)}{4a}t}(a_1+\mathrm{i}b_1)(a_2+\mathrm{i}b_2)}, \tag{76}$$

hence the function $|q(t,x)|^2$ is

$$|q(t,x)|^2 = \frac{\mathrm{e}^{\phi}}{1-2\gamma\mathrm{e}^{\theta}+\mathrm{e}^{2\theta}}, \tag{77}$$

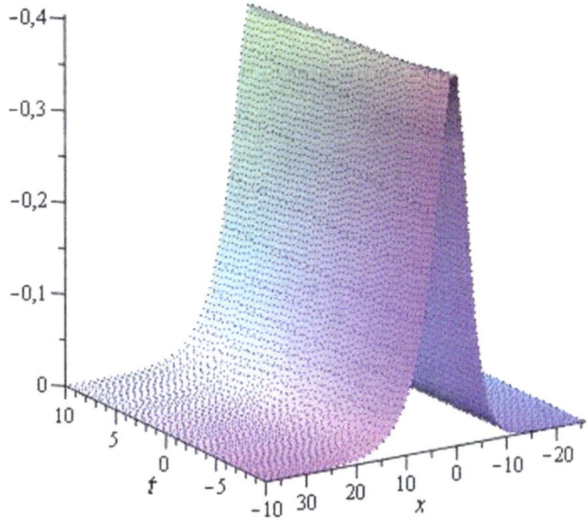

Fig. 3 An asymptotically decaying soliton corresponding to (76)

where

$$\theta = (k_1 + k_2)x - \frac{1}{4a}(k_1^3 + k_2^3)t,\ \phi = 2k_1 x - \frac{k_1^3}{2a}t,$$

$\gamma = (a_1 a_2 - b_1 b_2)/(k_1 + k_2)^2$ with $a_1^2 + b_1^2 = -(k_1 + k_2)^2/k$ and $a_2^2 + b_2^2 = -k(k_1 + k_2)^2$, $k_1 \neq -k_2$. Here $k_1, k_2 \in \mathbb{R}$. The above function has singularity when $e^{\theta} = \gamma \pm \sqrt{\gamma^2 - 1}$. Hence for $\gamma < 1$ and $k_1 > 0, k_2 > 0$ the solution is non-singular and bounded.

Example 3 For the set of the parameters $(k_1, k_2, e^{\delta_1}, e^{\delta_2}, k, a) = (\frac{1}{2}, \frac{1}{4}, -\frac{3}{4}, \frac{3}{4}, -1, 2)$ we obtain the following asymptotically decaying soliton

$$q(t,x) = \frac{(-3e^{\frac{1}{2}x - \frac{1}{64}t})}{4(1 + e^{\frac{3}{4}x - \frac{9}{512}t})},$$

whose graph is given in Fig. 3.

Remark 1 All dynamical variables considered so far are complex valued functions. We claim that all the results presented here will be valid if the dynamical variables are pseudo complex valued functions. Any pseudo complex number is $\alpha = a + ib$ where $i^2 = 1$. Complex conjugation is $\bar{\alpha} = a - ib$. Hence the norm of a pseudo complex number is not positive definite $\alpha\bar{\alpha} = a^2 - b^2$. NLS equation

$$iq_t = -\frac{1}{2}q_{xx} + kq^2\bar{q}, \tag{78}$$

has real and imaginary parts ($q = u + iv$)

$$u_t = -\frac{1}{2}v_{xx} + k(u^2 - \varepsilon v^2)v,$$
$$\varepsilon v_t = -\frac{1}{2}u_{xx} + k(u^2 - \varepsilon v^2)u,$$

where $i^2 = \varepsilon = \pm 1$.

5 Fordy–Kulish System

Systems of integrable nonlinear partial differential equations arise when the Lax pairs are given in certain Lie algebras. Fordy–Kulish (FK) system of equations are examples of such equations [9, 10]. We briefly give the Lax representations of these equations,

$$\phi_x = (\lambda H_S + Q^A E_A)\,\phi, \tag{79}$$
$$\phi_t = (A^a H_a + B^A E_A + C^D E_D)\,\phi, \tag{80}$$

where the dynamical variables are $Q^A = (q^\alpha, p^\alpha)$, the functions A^a, B^A, and C^D depend on the spectral parameter λ, on the dynamical variables (q^α, p^α) and their x-derivatives (for more details see [10, 15, 18]). The system of FK equations is an example when the functions A, B, and C are quadratic functions of λ. Let $q^\alpha(t, x)$ and $p^\alpha(t, x)$ be the complex dynamical variables where $\alpha = 1, 2, \ldots, N$, then the FK integrable system arising from the integrability condition of Lax equations (79) and (80) is given by

$$aq^\alpha_t = q^\alpha_{xx} + R^\alpha{}_{\beta\gamma-\delta}\, q^\beta\, q^\gamma\, p^\delta, \tag{81}$$
$$ap^\alpha_t = p^\alpha_{xx} + R^{-\alpha}{}_{-\beta-\gamma\delta}\, p^\beta\, p^\gamma\, q^\delta, \tag{82}$$

for all $\alpha = 1, 2, \ldots, N$. Here $R^\alpha{}_{\beta\gamma-\delta}$ and $R^{-\alpha}{}_{-\beta-\gamma\delta}$ are the curvature tensors of a Hermitian symmetric space satisfying

$$(R^\alpha{}_{\beta\gamma-\delta})^\star = R^{-\alpha}{}_{-\beta-\gamma\delta}, \tag{83}$$

and a is a complex number. These equations are known as the FK system which is integrable in the sense that they are obtained from the zero curvature condition of a connection defined on a Hermitian symmetric space and these equations can also be written in a Hamiltonian form.

The standard reduction of the above FK system is obtained by letting $p^\alpha = k(q^\alpha)^\star$ for all $\alpha = 1, 2, \ldots, N$. The FK system (81)–(82) reduces to a single equation

$$aq^\alpha_t = q^\alpha_{xx} + k\, R^\alpha{}_{\beta\gamma-\delta}\, q^\beta\, q^\gamma\, (q^\delta)^\star, \quad \alpha = 1, 2, \ldots, N, \tag{84}$$

provided that $a^\star = -a$ and (83) is satisfied. Here $*$ over a letter denotes complex conjugation.

6 Nonlocal Fordy–Kulish Equations

Here we will show that the Fordy–Kulish system is compatible with the nonlocal reduction of Ablowitz–Musslimani type. For this purpose using a similar constraint as in NLS system we let

$$p^\alpha(t,x) = k[q^\alpha(\mu_1 t, \mu_2 x)]^\star, \quad \alpha = 1, 2, \ldots, N, \tag{85}$$

where $\mu_1^2 = \mu_2^2 = 1$. Under this constraint the FK system (81)–(82) reduces to the following system of equations:

$$aq_t^\alpha(t,x) = q_{xx}^\alpha(t,x) + k\, R^\alpha{}_{\beta\gamma-\delta}\, q^\beta(t,x)\, q^\gamma(t,x)\, (q^\delta(\mu_1 t, \mu_2 x))^\star, \tag{86}$$

provided that $a^\star = -\mu_1 a$ and (83) is satisfied. In addition to (86) we have also an equation for $q^\delta(\mu_1 t, \mu_2 x)$ which can be obtained by letting $t \to \mu_1 t$, $x \to \mu_2 x$ in (86). Hence we obtain T-symmetric, S-symmetric, and ST-symmetric nonlocal FK equations. Nonlocal reductions correspond to $(\mu_1, \mu_2) = \{(-1, 1), (1, -1), (-1, -1)\}$. Hence corresponding to these values of μ_1 and μ_2 we have three different nonlocal integrable FK equations. They are given as follows:

1. T-Symmetric Nonlocal FK Equations:

$$aq_t^\alpha(t,x) = q_{xx}^\alpha(t,x) + k\, R^\alpha{}_{\beta\gamma-\delta}\, q^\beta(t,x)\, q^\gamma(t,x)\, (q^\delta(-t,x))^\star, \tag{87}$$

with $a^\star = a$.

2. S-Symmetric Nonlocal FK Equations:

$$aq_t^\alpha(t,x) = q_{xx}^\alpha(t,x) + k\, R^\alpha{}_{\beta\gamma-\delta}\, q^\beta(t,x)\, q^\gamma(t,x)\, (q^\delta(t,-x))^\star, \tag{88}$$

with $a^\star = -a$.

3. ST-Symmetric Nonlocal FK Equations:

$$aq_t^\alpha(t,x) = q_{xx}^\alpha(t,x) + k\, R^\alpha{}_{\beta\gamma-\delta}\, q^\beta(t,x)\, q^\gamma(t,x)\, (q^\delta(-t,-x))^\star, \tag{89}$$

with $a^\star = a$. All these three nonlocal equations are integrable.

7 Super Integrable Systems

When the Lax pair, in $(1+1)$-dimensions, is given in a super Lie algebra the resulting evolution equations are super integrable systems. They are given as a coupled system

$$q_t^i = F^i(q^k, \varepsilon^k, q_x^k, \varepsilon_x^k, q_{xx}^k, \varepsilon_{xx}^k, \ldots), \tag{90}$$

$$\varepsilon_t^i = G^i(q^k, \varepsilon^k, q_x^k, \varepsilon_x^k, q_{xx}^k, \varepsilon_{xx}^k, \ldots), \tag{91}$$

for all $i = 1, 2, \ldots, N$ where F^i and G^i $(i = 1, 2, \ldots, N)$ are functions of the dynamical variables $q^i(t, x)$, $\varepsilon^i(t, x)$, and their partial derivatives with respect to x. Here q^i's are bosonic and ε^i's are the fermionic dynamical variables. Since we start with a super Lax pair then the system (90)–(91) is a super integrable system of nonlinear partial differential equations.

8 Nonlocal Super NLS and MKdV Equations

As an example taking the Lax pair in super $sl(2, R)$ algebra we obtain the super AKNS system. We have two bosonic (q, r) and two fermionic (ε, β) dynamical variables. They satisfy the following evolution equations [16–18]:

i. Bosonic Equations

$$q_t = a_2\,(-\frac{1}{2}\,q_{xx} + q^2\,r + 2\,\varepsilon_x\,\varepsilon + 2q\beta\varepsilon) + \mathrm{i}a_3\,(-\frac{1}{4}\,q_{xxx} + \frac{3}{2}\,qrq_x + 3(\varepsilon_x\varepsilon)_x$$
$$-3q\beta_x\varepsilon + 3q\beta\varepsilon_x), \tag{92}$$

$$r_t = a_2\,(\frac{1}{2}\,r_{xx} - q\,r^2 + 2\,\beta_x\,\beta - 2r\beta\varepsilon) + \mathrm{i}a_3\,(-\frac{1}{4}\,r_{xxx} + \frac{3}{2}\,qrr_x - 3(\beta_x\beta)_x$$
$$+3r\beta_x\varepsilon - 3r\beta\varepsilon_x), \tag{93}$$

ii. Fermionic Equations

$$\beta_t = a_2\,(\beta_{xx} - r\varepsilon_x - \frac{1}{2}\,\varepsilon r_x - \frac{1}{2}qr\beta) + \mathrm{i}a_3(-\beta_{xxx} + \frac{3}{4}rq_x\beta + \frac{3}{4}qr_x\beta + \frac{3}{2}qr\beta_x$$
$$+\frac{3}{2}r_x\varepsilon_x + \frac{3}{4}\varepsilon r_{xx}), \tag{94}$$

$$\varepsilon_t = a_2\,(-\varepsilon_{xx} + q\beta_x + \frac{1}{2}\,\beta q_x + \frac{1}{2}qr\varepsilon) + \mathrm{i}a_3(-\varepsilon_{xxx} + \frac{3}{4}rq_x\varepsilon + \frac{3}{4}qr_x\varepsilon + \frac{3}{2}qr\varepsilon_x$$
$$+\frac{3}{2}q_x\beta_x + \frac{3}{4}\beta q_{xx}), \tag{95}$$

where a_2 and a_3 are arbitrary constants.

8.1 Super NLS Equations

Letting $a_3 = 0$ in the equations (92)–(95) we get the super coupled NLS system of equations. There are two bosonic (q,r) and two fermionic (ε, β) potentials satisfying

$$aq_t = -\frac{1}{2} q_{xx} + q^2 r + 2\varepsilon_x \varepsilon + 2q \beta \varepsilon, \tag{96}$$

$$ar_t = \frac{1}{2} r_{xx} - q r^2 + 2\beta_x \beta - 2r \beta \varepsilon, \tag{97}$$

$$a\varepsilon_t = -\varepsilon_{xx} + q \beta_x + \frac{1}{2} \beta q_x + \frac{1}{2} q r \varepsilon, \tag{98}$$

$$a\beta_t = \beta_{xx} - r \varepsilon_x - \frac{1}{2} \varepsilon r_x - \frac{1}{2} q r \beta, \tag{99}$$

where $a_2 = 1/a$. The standard reduction is $r = k_1 \bar{q}$ and $\beta = k_2 \bar{\varepsilon}$ where k_1 and k_2 are constants, a bar over a quantity denotes the Berezin conjugation in the Grassmann algebra. If P and Q are super functions (bosonic or fermionic) then $\overline{PQ} = \overline{Q}\,\overline{P}$. Under these constraints the above equations (96)–(99) reduce to the following super NLS equations provided $k_1 = k_2^2$ and $\bar{a} = -a$,

$$aq_t = -\frac{1}{2} q_{xx} + k_1 q^2 \bar{q} + 2\varepsilon_x \varepsilon + 2k_2 q \bar{\varepsilon} \varepsilon, \tag{100}$$

$$a\varepsilon_t = -\varepsilon_{xx} + k_2 q \bar{\varepsilon}_x + \frac{1}{2} k_2 \bar{\varepsilon} q_x + \frac{1}{2} k_1 q \bar{q} \varepsilon. \tag{101}$$

Here we show that super NLS system (96)–(99) can be reduced to nonlocal super NLS equations. This can be done by choosing the super Ablowitz–Musslimani reduction as

$$r(t,x) = k_1 \bar{q}(\mu_1 t, \mu_2 x), \quad \beta(t,x) = k_2 \bar{\varepsilon}(\mu_1 t, \mu_2 x). \tag{102}$$

where $\mu_1^2 = \mu_2^2 = 1$. Here k_1 and k_2 are real constants. Under these constraints the above set (96)–(99) reduces to super NLS equations [26, 27],

$$\begin{aligned} aq_t(t,x) &= -\frac{1}{2} q_{xx}(t,x) + k_1\, q^2(t,x)\, \bar{q}(\mu_1 t, \mu_2 x) + 2\varepsilon_x(t,x)\, \varepsilon(t,x) \\ &\quad + 2k_2\, q(t,x)\, \bar{\varepsilon}(\mu_1 t, \mu_2 x) \varepsilon(t,x), \\ a\varepsilon_t(t,x) &= -\varepsilon_{xx}(t,x) + k_2\, q(t,x)\, \bar{\varepsilon}_x(\mu_1 t, \mu_2 x) + \frac{1}{2} k_2\, \bar{\varepsilon}(\mu_1 t, \mu_2 x)\, q_x(t,x) \\ &\quad + \frac{1}{2} k_1\, q(t,x)\, \bar{q}(\mu_1 t, \mu_2 x)\, \varepsilon(t,x), \end{aligned}$$

provided that

$$\bar{a}\, \mu_1 = -a, \quad k_2^2\, \mu_2 = k_1. \tag{103}$$

Nonlocal reductions correspond to the choices $(\mu_1, \mu_2) = \{(-1, 1), (1, -1), (-1, -1)\}$. They are explicitly given by,

1. T-Symmetric Nonlocal Super NLS Equations:

$$\begin{aligned} aq_t(t,x) &= -\frac{1}{2} q_{xx}(t,x) + k_1\, q^2(t,x)\,\bar{q}(-t,x) + 2\varepsilon_x(t,x)\,\varepsilon(t,x) \\ &\quad +2k_2\, q(t,x)\,\bar{\varepsilon}(-t,x), \varepsilon(t,x), \\ a\varepsilon_t(t,x) &= -\varepsilon_{xx}(t,x) + k_2\, q(t,x)\,\bar{\varepsilon}_x(-t,x) + \frac{1}{2} k_2\, \bar{\varepsilon}(-t,x)\, q_x(t,x) \\ &\quad +\frac{1}{2} k_1\, q(t,x)\,\bar{q}(-t,x)\,\varepsilon(t,x), \end{aligned}$$

with $a^\star = a$ and $k_1 = k_2^2$.

2. S-Symmetric Nonlocal Super NLS Equations:

$$\begin{aligned} aq_t(t,x) &= -\frac{1}{2} q_{xx}(t,x) + k_1\, q^2(t,x)\,\bar{q}(t,-x) + 2\varepsilon_x(t,x)\,\varepsilon(t,x) + 2k_2\, q(t,x)\,\bar{\varepsilon}(t,-x)\varepsilon(t,x), \\ a\varepsilon_t(t,x) &= -\varepsilon_{xx}(t,x) + k_2\, q(t,x)\,\bar{\varepsilon}_x(t,-x) + \frac{1}{2} k_2\, \bar{\varepsilon}(t,-x)\, q_x(t,x) \\ &\quad +\frac{1}{2} k_1\, q(t,x)\,\bar{q}(t,-x)\,\varepsilon(t,x), \end{aligned}$$

with $a^\star = -a$ and $k_1 = -k_2^2$.

3. ST-Symmetric Nonlocal Super NLS Equations:

$$\begin{aligned} aq_t(t,x) &= -\frac{1}{2} q_{xx}(t,x) + k_1\, q^2(t,x)\,\bar{q}(-t,-x) + 2\varepsilon_x(t,x)\,\varepsilon(t,x) \\ &\quad +2k_2\, q(t,x)\,\bar{\varepsilon}(-t,-x), \varepsilon(t,x), \\ a\varepsilon_t(t,x) &= -\varepsilon_{xx}(t,x) + k_2\, q(t,x)\,\bar{\varepsilon}_x(-t,-x) + \frac{1}{2} k_2\, \bar{\varepsilon}(-t,-x)\, q_x(t,x) \\ &\quad +\frac{1}{2} k_1\, q(t,x)\,\bar{q}(-t,-x)\,\varepsilon(t,x), \end{aligned}$$

with $a^\star = a$ and $k_1 = -k_2^2$.

8.2 Super MKdV Systems

Another special case of the super AKNS equations is the super mKdV system [16, 17],

$$
\begin{aligned}
a q_t &= -\frac{1}{4} q_{xxx} + \frac{3}{2} r\, q\, q_x + 3(\varepsilon_x\, \varepsilon)_x - 3\, q\, \beta_x\, \varepsilon + 3 q \beta\, \varepsilon_x, \\
a r_t &= -\frac{1}{4} r_{xxx} + \frac{3}{2} r\, q\, r_x - 3(\beta_x\, \beta)_x + 3\, r\, \beta_x\, \varepsilon - 3 r \beta\, \varepsilon_x, \\
a \varepsilon_t &= -\varepsilon_{xxx} + \frac{3}{4} (r\, q)_x\, \varepsilon + \frac{3}{2} q\, r\, \varepsilon_x + \frac{3}{2} q_x\, \beta_x + \frac{3}{4} \beta\, q_{xx}, \\
a \beta_t &= -\beta_{xxx} + \frac{3}{4} (r\, q)_x\, \beta + \frac{3}{2} q\, r\, \beta_x + \frac{3}{2} r_x\, \varepsilon_x + \frac{3}{4} \varepsilon\, r_{xx}.
\end{aligned}
$$

The standard reduction is $r = k_1 \bar{q}$, $\beta = k_2\, \bar{\varepsilon}$. Then we obtain [16],

$$
\begin{aligned}
a q_t &= -\frac{1}{4} q_{xxx} + \frac{3}{2} k_1\, \bar{q}\, q\, q_x + 3(\varepsilon_x\, \varepsilon)_x - 3\, k_2 q\, \bar{\varepsilon}_x\, \varepsilon + 3 k_2\, q \bar{\varepsilon}\, \varepsilon_x, \\
a \varepsilon_t &= -\varepsilon_{xxx} + \frac{3}{4} k_1 (\bar{q}\, q)_x\, \varepsilon + \frac{3}{2} k_1 q\, \bar{q}\, \varepsilon_x + \frac{3}{2} k_2\, q_x\, \bar{\varepsilon}_x + \frac{3}{4} k_2 \bar{\varepsilon}\, q_{xx},
\end{aligned}
$$

provided that $k_1 = k_2^2$ and $\bar{a} = a$. For the super mKdV system, Ablowitz–Musslimani type of reduction is also possible. Letting

$$
r(t, x) = k_1\, \bar{q}(\mu_1 t, \mu_2 x), \quad \beta(t, x) = k_2 \bar{\varepsilon}(\mu_1 t, \mu_2 x), \tag{104}
$$

where $\mu_1^2 = \mu_2^2 = 1$ we get the following system of equations

$$
\begin{aligned}
a q_t(t, x) &= -\frac{1}{4} q_{xxx}(t, x) + \frac{3}{2} k_1\, \bar{q}(\mu_1 t, \mu_2 x)\, q(t, x)\, q_x(t, x) + 3(\varepsilon_x(t, x)\, \varepsilon(t, x))_x \\
&\quad -3\, q(t, x)\, \bar{\varepsilon}_x(\mu_1 t, \mu_2 x)\, \varepsilon(t, x) + 3 k_2\, q(t, x)\, \bar{\varepsilon}(\mu_1 t, \mu_2 x)\, \varepsilon_x(t, x), \qquad (105) \\
a \varepsilon_t(t, x) &= -\varepsilon_{xxx}(t, x) + \frac{3}{4} k_1 (\bar{q}(\mu_1 t, \mu_2 x)\, q(t, x))_x\, \varepsilon(t, x) + \frac{3}{2} k_1 q(t, x)\, \bar{q}(\mu_1 t, \mu_2 x)\, \varepsilon_x(t, x) \\
&\quad + \frac{3}{2} k_2\, q_x(t, x)\, \bar{\varepsilon}_x(\mu_1 t, \mu_2 x) + \frac{3}{4} k_2 \bar{\varepsilon}(\mu_1 t, \mu_2 x)\, q_{xx}(t, x), \qquad (106)
\end{aligned}
$$

provided that $\bar{a}\, \mu_1\, \mu_2 = a, k_2^2 \mu_2 = k_1$. Nonlocal reductions correspond to the choices $(\mu_1, \mu_2) = \{(-1, 1), (1, -1), (-1, -1)\}$. They are explicitly given by,

1. T-Symmetric Nonlocal Super MKdV Equations: Here $\bar{a} = -a$ and $k_1 = k_2^2$.

$$
\begin{aligned}
a q_t(t, x) &= -\frac{1}{4} q_{xxx}(t, x) + \frac{3}{2} k_1\, \bar{q}(-t, x)\, q(t, x)\, q_x(t, x) + 3(\varepsilon_x(t, x)\, \varepsilon(t, x))_x \\
&\quad -3\, q(t, x)\, \bar{\varepsilon}_x(-t, x)\, \varepsilon(t, x) + 3 k_2\, q(t, x) \bar{\varepsilon}(-t, x)\, \varepsilon_x(t, x), \qquad (107) \\
a \varepsilon_t(t, x) &= -\varepsilon_{xxx}(t, x) + \frac{3}{4} k_1 (\bar{q}(-t, x)\, q(t, x))_x\, \varepsilon(t, x) + \frac{3}{2} k_1 q(t, x)\, \bar{q}(-t, x)\, \varepsilon_x(t, x) \\
&\quad + \frac{3}{2} k_2\, q_x(t, x)\, \bar{\varepsilon}_x(-t, x) + \frac{3}{4} k_2 \bar{\varepsilon}(-t, x)\, q_{xx}(t, x), \qquad (108)
\end{aligned}
$$

2. S-Symmetric Nonlocal Super MKdV Equations: Here $\bar{a} = -a$ and $k_1 = -k_2^2$.

$$
\begin{aligned}
aq_t(t,x) &= -\frac{1}{4}\, q_{xxx}(t,x) + \frac{3}{2}\, k_1\, \bar{q}(t,-x)\, q(t,x)\, q_x(t,x) + 3(\varepsilon_x(t,x)\, \varepsilon(t,x))_x \\
&\quad -3\, q(t,x)\, \bar{\varepsilon}_x(t,-x)\, \varepsilon(t,x) + 3k_2\, q(t,x)\bar{\varepsilon}(t,-x)\, \varepsilon_x(t,x), \qquad (109)\\
a\varepsilon_t(t,x) &= -\varepsilon_{xxx}(t,x) + \frac{3}{4}\, k_1(\bar{q}(t,-x)\, q(t,x))_x\, \varepsilon(t,x) + \frac{3}{2}\, k_1 q(t,x)\, \bar{q}(t,-x)\, \varepsilon_x(t,x) \\
&\quad +\frac{3}{2}\, k_2\, q_x(t,x)\, \bar{\varepsilon}_x(t,-x) + \frac{3}{4}\, k_2 \bar{\varepsilon}(t,-x)\, q_{xx}(t,x), \qquad (110)
\end{aligned}
$$

3. ST-Symmetric Nonlocal Super MKdV Equations: Here $\bar{a} = a$ and $k_1 = -k_2^2$.

$$
\begin{aligned}
aq_t(t,x) &= -\frac{1}{4}\, q_{xxx}(t,x) + \frac{3}{2}\, k_1\, \bar{q}(-t,-x)\, q(t,x)\, q_x(t,x) + 3(\varepsilon_x(t,x)\, \varepsilon(t,x))_x \\
&\quad -3\, q(t,x)\, \bar{\varepsilon}_x(-t,-x)\, \varepsilon(t,x) + 3k_2\, q(t,x)\bar{\varepsilon}(-t,-x)\, \varepsilon_x(t,x), \qquad (111)\\
a\varepsilon_t(t,x) &= -\varepsilon_{xxx}(t,x) + \frac{3}{4}\, k_1(\bar{q}(-t,-x)\, q(t,x))_x\, \varepsilon(t,x) + \frac{3}{2}\, k_1 q(t,x)\, \bar{q}(-t,-x)\, \varepsilon_x(t,x) \\
&\quad +\frac{3}{2}\, k_2\, q_x(t,x)\, \bar{\varepsilon}_x(-t,-x) + \frac{3}{4}\, k_2 \bar{\varepsilon}(-t,-x)\, q_{xx}(t,x). \qquad (112)
\end{aligned}
$$

9 Concluding Remarks

In this work we first presented all integrable nonlocal reductions of NLS and mKdV systems. We gave the recursion operators and the soliton solutions of these nonlocal equations. We then presented the extension of the nonlocal NLS equation to nonlocal Fordy–Kulish equations on symmetric spaces. Starting with the super AKNS system we studied all possible nonlocal reductions and found two new super integrable systems. They are the nonlocal super NLS equations and nonlocal super mKdV systems of equations. There are three different nonlocal types of super integrable equations. They correspond to T-, S-, and ST- symmetric super NLS and super mKdV equations.

From the study of NLS and mKdV systems (both bosonic and fermionic integrable systems) we observed that they have standard and nonlocal reductions. Moreover in both of these systems there are at least one nonlocal reduction to a standard reduction. For instance both systems have $r(t,x) = k\bar{q}(t,x)$ as a standard reduction and the corresponding nonlocal reductions are $r(t,x) = k\bar{q}(\mu_1 t, \mu_2 x)$ where k is real constant and $(\mu_1, \mu_2) = \{(1,-1), (-1,1), (-1,-1)\}$. From these reductions we obtain standard and nonlocal NLS equations and standard and nonlocal complex mKdV equations and their nonlocal super integrable extensions. The mKdV system has additional standard and nonlocal reductions. Standard reduction is $r(t,x) = kq(t,x)$, where k is real constant, and its corresponding nonlocal reduction $r(t,x) = kq(-t,-x)$ gives the nonlocal mKdV equation. From all these experiences we conclude with a conjecture: If a system of equations admits a standard reduction then there exists at least one corresponding nonlocal reduction of the same system.

Acknowledgements This work is partially supported by the Scientific and Technological Research Council of Turkey (TÜBİTAK).

References

1. Ablowitz, M.J., Musslimani, Z.H.: Integrable nonlocal nonlinear Schrödinger equation. Phys. Rev. Lett. **110**, 064105 (2013)
2. Ablowitz, M.J., Musslimani, Z.H.: Inverse scattering transform for the integrable nonlocal nonlinear Schrödinger equation. Nonlinearity **29**, 915–946 (2016)
3. Ablowitz, M.J., Musslimani, Z.H.: Integrable nonlocal nonlinear equations. Stud. Appl. Math. **139**(1), 7–59 (2016)
4. Ablowitz, M.J., Kaup, D.J., Newell, A.C., Segur, H.: The inverse scattering transform-fourier analysis for nonlinear problems. Stud. Appl. Math. **53**(4), 249–315 (1974)
5. Caudrelier, V.: Interplay between the inverse scattering method and Fokas's unified transform with an application. Stud. App. Math. **140**, 3–26 (2017). arXiv:1704.05306v4 [math-ph]
6. Chen, K., Deng, X., Lou, S., Zhang, D.: Solutions of nonlocal equations reduced from the AKNS hierarchy. Stud. App. Math. (2018) (to appear). arXiv:1710.10479 [nlin.SI]
7. Feng, B.F., Luo, X.D., Ablowitz, M.J., Musslimani, Z.H.: General soliton solution to a nonlocal nonlinear Schrödinger equation with zero and nonzero boundary conditions. arXiv:1712.09172 [nlin.SI]
8. Fokas, A.S.: Integrable multidimensional versions of the nonlocal Schrödinger equation. Nonlinearity **29**, 319–324 (2016)
9. Fordy, A.P.: Derivative nonlinear Schrödinger equations and hermitian symmetric spaces. J. Phys. A Math. Gen. **17**, 1235–1245 (1984)
10. Fordy, A.P., Kulish, P.P.: Nonlinear Schrödinger equations and simple Lie algebras. Commun. Math. Phys. **89**, 427–443 (1983)
11. Gerdjikov, V.S.: On nonlocal models of Kulish-Sklyanin type and generalized fourier transforms. Stud. Comp. Int. **681**, 37–52 (2017). arXiv:1703.03705 [nlin.SI]
12. Gerdjikov, V.S., Saxena, A.: Complete integrability of nonlocal nonlinear Schrödinger equation. J. Math. Phys. **58**(1), 013502 (2017). arXiv:1510.00480 [nlin.SI]
13. Gerdjikov, V.S., Grahovski, D.G., Ivanov, R.I.: On the N-wave equations with PT symmetry. Theor. Math. Phys. **188**(3), 1305–1321 (2016)
14. Gerdjikov, V.S., Grahovski, D.G., Ivanov, R.I.: On the integrable wave interactions and Lax pairs on symmetric spaces. Wave Motion **71**, 53–70 (2017)
15. Gürses, M.: Nonlocal Fordy-Kulish equations on symmetric spaces. Phys. Lett. A **381**, 1791–1794 (2017)
16. Gürses, M., Oğuz, Ö.: A super AKNS scheme. Phys. Lett. A **108**(9), 437–440 (1985)
17. Gürses, M., Oğuz, Ö.: A super soliton connection. Lett. Math. Phys. **11**, 235–246 (1986)
18. Gürses, M., Oğuz, Ö., Salihoğlu, S.: Nonlinear partial differential equations on homogeneous spaces. Int. J. Mod. Phys. A **5**, 1801–1817 (1990)
19. Gürses, M., Pekcan, A.: Nonlocal nonlinear Schrödinger equations and their soliton solutions. J. Math. Phys. **59**, 051501 (2018). arXiv:1707.07610v1 [nlin.SI]
20. Gürses, M., Pekcan, A.: Nonlocal nonlinear modified KdV equations and their soliton solutions. Commun. Nonlinear Sci. Numer. Simul. **67**, 427–448 (2019). arXiv:1711.01588 [nlin.SI]
21. Huang, X., King, L.: Soliton solutions for the nonlocal nonlinear Schrödinger equation. Eur. Phys. J. Plus **131**, 148 (2016)
22. Iwao, M., Hirota, R.: Soliton solutions of a coupled modified KdV equations. J. Phys. Soc. Jpn. **66**(3), 577–588 (1997)
23. Ji, J.L., Zhu, Z.N.: On a nonlocal modified Korteweg-de Vries equation: integrability, Darboux transformation and soliton solutions. Commun. Nonlinear Sci. Numer. Simul. **42**, 699–708 (2017)

24. Ji, J.L., Zhu, Z.N.: Soliton solutions of an integrable nonlocal modified Korteweg-de Vries equation through inverse scattering transform. J. Math. Anal. Appl. **453**, 973–984 (2017). arXiv:1603.03994 [nlin.SI]
25. Khare, A., Saxena, A.: Periodic and hyperbolic soliton solutions of a number of nonlocal nonlinear equations. J. Math. Phys. **56**, 032104 (2015)
26. Kupperschmidt, B.A.: A super Korteweg-de Vries equation: an integrable system. Phys. Lett. **102A**, 213 (1983)
27. Kupperschmidt, B.A.: Bosons and Fermions interacting integrably with the Korteweg-de Vries field. J. Phys. A Math. Gen. **17**, L869 (1984)
28. Li, M., Xu, T.: Dark and antidark soliton interactions in the nonlocal nonlinear Schrödinger equation with the self-induced parity-time-symmetric potential. Phys. Rev. E **91**, 033202 (2015)
29. Ma, L.Y., Shen, S.F., Zhu, Z.N.: Integrable nonlocal complex mKdV equation: soliton solution and Gauge equivalence. arXiv:1612.06723 [nlin.SI]
30. Sakkaravarthi, K., Kanna, T.: Bright solitons in coherently coupled nonlinear Schrödinger equations with alternate signs of nonlinearities. J. Math. Phys. **54**, 013701 (2013)
31. Sinha, D., Ghosh, P.K.: Integrable nonlocal vector nonlinear Schrödinger equation with self-induced parity-time symmetric potential. Phys. Lett. A **381**, 124–128 (2017)
32. Wen, X.Y., Yan, Z., Yang, Y.: Dynamics of higher-order rational solitons for the nonlocal nonlinear Schrödinger equation with the self-induced parity-time-symmetric potential. Chaos **26**, 063123 (2015)
33. Yang, J.: General N-solitons and their dynamics in several nonlocal nonlinear Schrödinger equations. arXiv:1712.01181 [nlin.SI]
34. Yang, B., Yang, J.: Transformations between nonlocal and local integrable equations. Stud. App. Math. **140**, 178–201 (2017). arXiv:1705.00332v1 [nlin.PS]

Construction of Solvable Structures from $\mathfrak{so}(3, \mathbb{C})$

A. Ruiz and C. Muriel

Abstract For third-order ordinary differential equations admitting a Lie symmetry algebra isomorphic to $\mathfrak{so}(3, \mathbb{C})$ it is proved the existence of a solvable structure constructed out of the symmetry generators of the algebra. This solvable structure is explicitly obtained in terms of solutions to a second-order linear ordinary differential equation. Once the solvable structure is known, a complete set of first integrals can be computed by quadratures. Moreover, it is proved that the general solution can be obtained in parametric form and expressed in terms of solutions to a second-order linear equation.

Keywords First integral · Solvable structure · Non-solvable symmetry algebra

1 Introduction

The knowledge of a solvable n-dimensional Lie symmetry algebra for an nth-order ordinary differential equation (ODE) permits to stepwise reduce the order of the equation and obtain the general solution after n successive quadratures [10, 11, 17, 18, 23]. However, if the Lie symmetry algebra is non-solvable then this step by step reduction method is no longer applicable because one of the symmetry generators is lost at some stage of the procedure. The first case of nth-order ODE admitting a non-solvable symmetry algebra occurs for $n = 3$, and corresponds to symmetry algebras isomorphic to either $\mathfrak{sl}(2, \mathbb{R})$ or $\mathfrak{so}(3, \mathbb{R})$. The case of $\mathfrak{sl}(2, \mathbb{R})$ was addressed in [15] by using techniques based on λ-symmetries [14] and it was proved that the Lie point symmetries that are lost in the reduction method can be recovered as λ-symmetries of the corresponding reduced equation. On the other hand, in [16] it was proved

A. Ruiz (✉) · C. Muriel (✉)
Department of Mathematics, University of Cádiz, 11510 Puerto Real, Spain
e-mail: adrian.ruiz@uca.es

C. Muriel
e-mail: concepcion.muriel@uca.es

V. G. Kac et al. (eds.), *Symmetries, Differential Equations and Applications*, Springer Proceedings in Mathematics & Statistics 266,
https://doi.org/10.1007/978-3-030-01376-9_3

that a convenient combination of the symmetry generators of $\mathfrak{so}(3, \mathbb{R})$ allows one to recover again the lost symmetries as λ-symmetries.

Other approaches that can be found in the recent literature as a generalization of the classical Lie method are hidden symmetries [1], μ-symmetries [6, 8], nonlocal symmetries [2] and solvable structures [3–5, 19]. In this paper we focus on the concept of solvable structures, which were introduced by Basarab-Horwath in [4] and further studied in [3, 5, 19]. Solvable structures generalize the classical result on the integrability by quadratures of involutive distributions of vector fields admitting a sufficiently large solvable symmetry algebra. In contrast, vector fields that are involved in a solvable structure are not needed to be symmetries of the distribution. This degree of freedom permits to characterize the integrability by quadratures by means of the existence of solvable structures. When the distribution is formed just by the vector field associated to an nth-order ODE, then the computation of a solvable structure warrants locally the integrability by quadratures of the given ODE, even if it does not admit a solvable n-dimensional symmetry algebra.

Although solvable structures are a powerful tool for the integrability by quadratures of ODEs, finding an explicit expression for a solvable structure associated to an nth-order equation is a difficult task in practice. One of the possible approaches to construct solvable structures may be to exploit the Lie symmetry algebra admitted by the equation. The case of $\mathfrak{sl}(2, \mathbb{R})$ has been recently solved in [20] (see also [21]) and solvable structures were explicitly computed by using the symmetry generators of the algebra. Those theoretical results were applied in [22] to compute first integrals and parametric general solutions in terms of solutions to second-order linear equations for third-order ODEs admitting a Lie symmetry algebra isomorphic to $\mathfrak{sl}(2, \mathbb{R})$.

The goal of this paper is to construct explicitly a solvable structure for third-order ODEs admitting the non-solvable symmetry algebra $\mathfrak{so}(3, \mathbb{C})$. Once the solvable structure is computed, we are able to obtain a complete set of first integrals expressed in terms of the solutions to a second-order linear equation. Furthermore, from the corresponding implicit solution, the general solution of the equation is obtained in parametric form, also in terms of the solutions to a related second-order linear equation. Other approaches to integrate ODEs admitting non-solvable symmetry algebras can be found in [7, 9, 12, 16].

The paper is organized as follows. In Sect. 2 we set up the notation and the definition of solvable structure as well as some basic properties adapted to the case of ODEs. In Sect. 3 solvable structures for third-order ODEs admitting $\mathfrak{so}(3, \mathbb{C})$ are constructed from the generators of the algebra by exploiting some results presented in [20, 21].

Once a procedure to compute solvable structures has been established, in Sect. 4 a complete set of first integrals for third-order ODEs admitting the Lie symmetry algebra $\mathfrak{so}(3, \mathbb{C})$ is computed in terms of a fundamental set of solutions to a second-order linear equation. Besides, from the corresponding implicit solution, the general solution of the equation is obtained in parametric form and expressed in terms of the solutions to a related second-order linear equation.

Finally, in Sect. 5 our method is illustrated with two examples of equations admitting a three-dimensional Lie symmetry algebra isomorphic to $\mathfrak{so}(3, \mathbb{C})$. Therefore the equations do not posses additional Lie point symmetries. The general solutions of the equations considered in Example I and Example II are obtained in parametric form and expressed in terms of the Kummer functions and the modified Bessel functions, respectively.

2 Solvable Structures for ODEs

In this section we recall the definition of solvable structure as well as its application to integrate ODEs by quadratures. For a more extensive study the reader is referred to [3–5, 19]. From this point on, functions and vector fields are assumed to be smooth and well defined on an open and simply connected subset D of either $\mathbb{R}^n$ or an n-dimensional manifold $\mathscr{M}_n$.

Definition 1 ([20, *Def. 2.1*]) Let $\mathscr{A} = \{\mathbf{A}_1, \ldots, \mathbf{A}_r\}$ be a system of $r < n$ independent vector fields on D which are in involution.

1. A smooth vector field $\mathbf{X}$ on D is called a *symmetry* of $\mathscr{A}$ if the following conditions hold:
   a. $\mathbf{A}_1, \ldots, \mathbf{A}_r$, and $\mathbf{X}$ are independent;
   b. $[\mathbf{X}, \mathbf{A}_i] \in \text{span}(\mathscr{A})$, for $1 \leq i \leq r$.
2. Let $\mathscr{S} = \langle \mathbf{X}_1, \ldots, \mathbf{X}_{n-r} \rangle$ be an ordered set of independent vector fields on D. The ordered system $\mathscr{A} \cup \mathscr{S} = \langle \mathbf{A}_1, \ldots, \mathbf{A}_r, \mathbf{X}_1, \ldots, \mathbf{X}_{n-r} \rangle$ is a *solvable structure* with respect to $\mathscr{A}$ if
   a. $\mathscr{S}_j = \{\mathbf{A}_1, \ldots, \mathbf{A}_r, \mathbf{X}_1, \ldots, \mathbf{X}_j\}$ is in involution, for $j = 1, \ldots, n-r$;
   b. $\mathbf{X}_1$ is a symmetry of $\mathscr{A}$;
   c. $\mathbf{X}_{j+1}$ is a symmetry of $\mathscr{S}_j$, for $j = 1, \ldots, n-r-1$.

The integrability by quadratures can be characterized by means of solvable structures:

Proposition 1 ([4, Prop. 6]) *An involutive system $\mathscr{A}$ is locally integrable by quadratures if and only if there exists a solvable structure with respect to $\mathscr{A}$.*

Solvable structures provide a systematic method to obtain a set of functionally independent first integrals $\{I_1, \ldots, I_{n-r}\}$ common to the system of vector fields $\mathscr{A}$, based on the integration by quadratures of $n-r$ one forms which have particular closure properties. The reader can consult [3–5, 19] for a deep study on this method.

With the goal of applying solvable structures in the context of ODEs, let us recall that for a given ODE of the form

$$\theta_n = F(r, \theta, \theta_1, \ldots, \theta_{n-1}), \tag{1}$$

where r denotes the independent variable, θ is the dependent variable and $\theta_j = \frac{d^j\theta}{dr^j}$ for $j = 1, \ldots, n$, the vector field $\mathbf{A}$ associated to Eq. (1) is given by

$$\mathbf{A} = \partial_r + \theta_1 \partial_{\theta_1} + \cdots + F(r, \theta, \theta_1, \ldots, \theta_{n-1}) \partial_{u_{n-1}}. \tag{2}$$

Therefore, if we consider the (trivially) involutive system of vector fields $\mathscr{A}$ formed by just the vector field (2) associated to an nth-order ODE (1), then Proposition 1 provides the following corollary:

Corollary 1 *An nth-order ODE (1) is locally integrable by quadratures if and only if there exists a solvable structure with respect to the vector field (2) associated to the equation.*

3 Solvable Structures from $\mathfrak{so}(3, \mathbb{C})$ for Third-Order ODEs

Let us consider a third-order ODE

$$\theta_3 = \varphi(r, \theta, \theta_1, \theta_2), \tag{3}$$

where $\theta_j = \frac{d^j\theta}{dr^j}$ for $1 \leq j \leq 3$, and let $M \subset \mathbb{C}^2$ be an open set of the projection of the domain of φ to the corresponding zero-order jet space. We denote by $\mathbf{A}_{(r,\theta)} = \partial_r + \theta_1 \partial_\theta + \theta_2 \partial_{\theta_1} + \varphi(r, \theta, \theta_1, \theta_2)\partial_{\theta_2}$ the vector field associated to Eq. (3) and suppose that Eq. (3) admits the Lie symmetry algebra $\mathfrak{so}(3, \mathbb{C})$.

A basis of generators $\{\mathbf{u}_1, \mathbf{u}_2, \mathbf{u}_3\}$ of the algebra satisfying the following commutation relations can be chosen [13]:

$$[\mathbf{u}_1, \mathbf{u}_2] = \mathbf{u}_3, \quad [\mathbf{u}_1, \mathbf{u}_3] = -\mathbf{u}_2, \quad [\mathbf{u}_2, \mathbf{u}_3] = \mathbf{u}_1. \tag{4}$$

The commutation relations (4) show that the Lie symmetry algebra $\mathfrak{so}(3, \mathbb{C})$ is non-solvable, therefore the Lie reduction method cannot be applied to stepwise reduce the order of Eq. (3). Furthermore, if we use any of the Lie point symmetries $\mathbf{u}_i$, for $i = 1, 2, 3$, to reduce the order of Eq. (3), then the remaining symmetry generators are lost as Lie point symmetries of the corresponding reduced equation.

Our goal in this section is to construct a solvable structure with respect to $\mathbf{A}_{(r,\theta)}$ from the basis elements of $\mathfrak{so}(3, \mathbb{C})$. By using the properties of the Lie bracket it can be checked that the vector fields

$$\mathbf{v}_1 = \mathbf{u}_2 - i\mathbf{u}_3, \quad \mathbf{v}_2 = \mathbf{u}_2 + i\mathbf{u}_3, \quad \mathbf{v}_3 = i\mathbf{u}_1, \tag{5}$$

satisfy the commutation relations:

$$[\mathbf{v}_1, \mathbf{v}_3] = \mathbf{v}_1, \quad [\mathbf{v}_1, \mathbf{v}_2] = 2\mathbf{v}_3, \quad [\mathbf{v}_3, \mathbf{v}_2] = \mathbf{v}_2. \tag{6}$$

Observe that, by (6), the system of vector fields $\left\{\mathbf{A}_{(r,\theta)}, \mathbf{v}_i^{(2)}, \mathbf{v}_3^{(2)}\right\}$ is involutive for $i = 1, 2$, hence, by Frobenius' Theorem, there exists a non-constant function $I_i = I_i(r, \theta, \theta_1, \theta_2)$ such that

$$\begin{cases} \mathbf{A}_{(r,\theta)}(I_i) = 0, \\ \mathbf{v}_i^{(2)}(I_i) \ = 0, \\ \mathbf{v}_3^{(2)}(I_i) \ = 0. \end{cases} \tag{7}$$

Conditions (7) imply that $\mathbf{v}_2^{(2)}(I_1) \neq 0$ and $\mathbf{v}_1^{(2)}(I_2) \neq 0$ because $\mathbf{A}_{(r,\theta)}$, $\mathbf{v}_1^{(2)}$, $\mathbf{v}_2^{(2)}$ and $\mathbf{v}_3^{(2)}$ are pointwise linearly independent vector fields defined on the four dimensional space of variables $(r, \theta, \theta_1, \theta_2)$. Therefore the functions

$$F_1 = \frac{1}{\mathbf{v}_1^{(2)}(I_2)} \quad \text{and} \quad F_2 = \frac{1}{\mathbf{v}_2^{(2)}(I_1)} \tag{8}$$

are locally well defined. Besides, it can be checked that F_1 and F_2 satisfy the following conditions [20, Lemma 4.2]:

$$\begin{cases} \mathbf{v}_3^{(2)}(F_1) \ = F_1, \\ \mathbf{A}_{(r,\theta)}(F_1) = 0, \\ \mathbf{v}_2^{(2)}(F_1) \ = 0, \end{cases} \qquad \begin{cases} \mathbf{v}_3^{(2)}(F_2) \ = -F_2, \\ \mathbf{A}_{(r,\theta)}(F_2) = 0, \\ \mathbf{v}_1^{(2)}(F_1) \ = 0, \end{cases} \tag{9}$$

and, in consequence, by using (5), the following lemma is proved:

Lemma 1 *The functions F_1 and F_2 defined in (8) satisfy*

$$\begin{cases} \mathbf{u}_1^{(2)}(F_1) \ = -i F_1, \\ \mathbf{A}_{(r,\theta)}(F_1) = 0, \\ \mathbf{u}_2^{(2)}(F_1) \ = -i\mathbf{u}_3^{(2)}(F_1), \end{cases} \qquad \begin{cases} \mathbf{u}_1^{(2)}(F_2) \ = i F_2, \\ \mathbf{A}_{(r,\theta)}(F_2) = 0, \\ \mathbf{u}_2^{(2)}(F_2) \ = i\mathbf{u}_3^{(2)}(F_2). \end{cases} \tag{10}$$

The compatibility of the systems given in (10) can be used to prove that both ordered sets $\langle \mathbf{A}_{(r,\theta)}, i\mathbf{u}_1^{(2)}, F_1(\mathbf{u}_2^{(2)} - i\mathbf{u}_3^{(2)}), F_2(\mathbf{u}_2^{(2)} + i\mathbf{u}_3^{(2)})\rangle$ and $\langle \mathbf{A}_{(r,\theta)}, i\mathbf{u}_1^{(2)}, F_2(\mathbf{u}_2^{(2)} + i\mathbf{u}_3^{(2)}), F_1(\mathbf{u}_2^{(2)} - i\mathbf{u}_3^{(2)})\rangle$ are solvable structures with respect to $\mathbf{A}_{(r,\theta)}$, as spelled out in the following theorem:

Theorem 1 *Let $\{\mathbf{u}_1, \mathbf{u}_2, \mathbf{u}_2\}$ be a basis of generators of $\mathfrak{so}(3, \mathbb{C})$ satisfying the commutation relations (4) and let F_1 and F_2 be two functions satisfying (10). Then, both ordered sets*

$$\langle \mathbf{A}_{(r,\theta)}, i\mathbf{u}_1^{(2)}, F_1(\mathbf{u}_2^{(2)} - i\mathbf{u}_3^{(2)}), F_2(\mathbf{u}_2^{(2)} + i\mathbf{u}_3^{(2)})\rangle \tag{11}$$

and

$$\langle \mathbf{A}_{(r,\theta)}, i\mathbf{u}_1^{(2)}, F_2(\mathbf{u}_2^{(2)} + i\mathbf{u}_3^{(2)}), F_1(\mathbf{u}_2^{(2)} - i\mathbf{u}_3^{(2)})\rangle \tag{12}$$

are solvable structures with respect to $\mathbf{A}_{(r,\theta)}$.

Proof Since it is assumed that Eq. (3) admits the Lie symmetry algebra $\mathfrak{so}(3, \mathbb{C})$, we have that

$$[\mathbf{u}_i^{(2)}, \mathbf{A}_{(r,\theta)}] = \rho_i \, \mathbf{A}_{(r,\theta)}, \quad \rho_i \in \mathscr{C}^\infty(M) \quad \text{for } i = 1, 2, 3. \tag{13}$$

It can be checked by using (4), (13) and Lemma 1 that the following commutation relations hold:

$$[\mathbf{A}_{(r,\theta)}, i\mathbf{u}_1^{(2)}] = -i\rho_1\mathbf{A}_{(r,\theta)}, \quad [\mathbf{A}_{(r,\theta)}, F_1(\mathbf{u}_2^{(2)} - i\mathbf{u}_3^{(2)})] = F_1(i\rho_3 - \rho_2)\,\mathbf{A}_{(r,\theta)}$$

$$[\mathbf{A}_{(r,\theta)}, F_2(\mathbf{u}_2^{(2)} + i\mathbf{u}_3^{(2)})] = -F_2(\rho_2 + i\rho_3)\mathbf{A}_{(r,\theta)},$$

$$[i\mathbf{u}_1^{(2)}, F_1(\mathbf{u}_2^{(2)} - i\mathbf{u}_3^{(2)})] = [i\mathbf{u}_1^{(2)}, F_2(\mathbf{u}_2^{(2)} + i\mathbf{u}_3^{(2)})] = 0,$$

$$[F_1(\mathbf{u}_2^{(2)} - i\mathbf{u}_3^{(2)}), F_2(\mathbf{u}_2^{(2)} + i\mathbf{u}_3^{(2)})] = 2F_1F_2i\mathbf{u}_1^{(2)},$$

which prove that (11) and (12) are solvable structures with respect to $\mathbf{A}_{(r,\theta)}$.

4 First Integrals and Parametric General Solution

Once either the solvable structure (11) or (12) is explicitly computed, a complete set of first integrals to Eq. (3) can be obtained by following the method presented in [4] (see also [3, 5, 19]). However, let us observe that both functions F_1 and F_2 defined in (8) are necessary for the construction of the solvable structures (11) and (12) and, by Lemma 1, such functions are already first integrals of the original equation. Furthermore, it can be checked that both sets $\{I_1, I_2, F_1\}$ and $\{I_1, I_2, F_2\}$ are complete sets of first integrals to Eq. (3) [20]. Therefore, in this section we focus on obtaining an explicit expression for the functions I_1, I_2 and F_1.

The action of the Lie group SO(3, $\mathbb{C}$), with Lie algebra $\mathfrak{so}(3, \mathbb{C})$, on any two-dimensional complex manifold can be modelled by the Lie algebra spanned by the vector fields [13]:

$$\begin{cases} \mathbf{u}_1 = \partial_\theta, \\ \mathbf{u}_2 = \dfrac{1}{2}(1+r^2)\cos\theta\,\partial_r + \dfrac{1}{2}(r - r^{-1})\sin\theta\,\partial_\theta, \\ \mathbf{u}_3 = -\dfrac{1}{2}(1+r^2)\sin\theta\,\partial_r + \dfrac{1}{2}(r - r^{-1})\cos\theta\,\partial_\theta. \end{cases} \tag{14}$$

The corresponding vector fields given in (5) become

$$\begin{cases} \mathbf{v}_1 = \dfrac{1}{2}(1+r^2)e^{i\theta}\,\partial_r - \dfrac{i}{2}(r - r^{-1})e^{i\theta}\,\partial_\theta, \\ \mathbf{v}_2 = \dfrac{1}{2}(1+r^2)e^{-i\theta}\,\partial_r + \dfrac{i}{2}(r - r^{-1})e^{-i\theta}\,\partial_\theta, \\ \mathbf{v}_3 = i\,\partial_\theta. \end{cases} \tag{15}$$

It can be checked that by means of the local change of variables

$$x = -\frac{1}{r}e^{-i\theta}, \quad u = re^{-i\theta}, \tag{16}$$

the symmetry generators (15) are respectively mapped into

$$\mathbf{v}_1 = \partial_x + \partial_u, \quad \mathbf{v}_2 = x^2\partial_x + u^2\partial_u, \quad \mathbf{v}_3 = x\partial_x + u\partial_u, \tag{17}$$

which satisfy the commutation relations (6). In fact, the vector fields (17) correspond with the basis elements of one of the canonical representations of $\mathfrak{sl}(2, \mathbb{C})$ considered in [22, Case 3 in Table 1]. The most general third-order ODE admitting the Lie symmetry algebra $\mathfrak{sl}(2, \mathbb{C})$ spanned by (17) is [22, Case 3 in Table 2]:

$$u_3 = \frac{3u_2^2}{2u_1} - \frac{u_1^2}{2(u-x)^2C(s)}, \quad \text{where } s = (2u_1 + 2u_1^2 + u_2(-u+x))u_1^{-3/2}. \tag{18}$$

A complete set of first integrals $\{I_1, I_2, F_1\}$ for a SL(2)-invariant third-order equation of the form (18) was reported in [22]: If ψ_1 and ψ_2 are two linearly independent solutions to the second-order linear equation

$$\psi''(s) + \frac{s\,C(s)^2 - C'(s)}{C(s)}\psi'(s) + 4C(s)^2\psi(s) = 0, \tag{19}$$

then a complete set of first integral for Eq. (18) is given by

$$I_1 = \frac{2\sqrt{u_1}C(s)\psi_1(s) + \psi_1'(s)}{2\sqrt{u_1}C(s)\psi_2(s) + \psi_2'(s)}, \quad I_2 = \frac{2\sqrt{u_1}x\,C(s)\psi_1(s) + u\psi_1'(s)}{2\sqrt{u_1}x\,C(s)\psi_2(s) + u\psi_2'(s)}, \tag{20}$$

$$F_1 = \frac{\left(2C(s)x\sqrt{u_1}\psi_2(s) + u\psi_2'(s)\right)^2}{2\sqrt{u_1}C(s)(u-x)W(\psi_1, \psi_2)(s)},$$

where $W(\psi_1, \psi_2)$ denotes the Wronskian determinant of ψ_1 and ψ_2. By expressing the first integrals given in (20) in terms of the original coordinates $\{r, \theta, \theta_1, \theta_2\}$ we obtain a complete set of first integrals for the original SO(3)-invariant third-order Eq. (3). The expressions of such first integrals are presented in the following theorem:

Theorem 2 *A complete set of first integrals for a third-order ODE admitting the Lie symmetry algebra* $\mathfrak{so}(3)$ *spanned by (14) is given by*

$$I_1 = \frac{2r\sqrt{1 + r^2\theta_1^2}C(\widehat{s})\psi_1(\widehat{s}) + (i\theta_1 r + 1)\psi_1'(\widehat{s})}{2r\sqrt{1 + r^2\theta_1^2}C(\widehat{s})\psi_2(\widehat{s}) + (i\theta_1 r + 1)\psi_2'(\widehat{s})},$$

$$I_2 = \frac{2\sqrt{1 + r^2\theta_1^2}C(\widehat{s})\psi_1(\widehat{s}) - (r + ir^2\theta_1)\psi_1'(\widehat{s})}{2\sqrt{1 + r^2\theta_1^2}C(\widehat{s})\psi_2(\widehat{s}) - (r + ir^2\theta_1)\psi_2'(\widehat{s})},$$

$$F_1 = \frac{e^{-i\theta}\left(2\sqrt{1+r^2\theta_1^2}C(\widehat{s})\psi_2(\widehat{s}) - (r+ir^2\theta_1)\psi_2'(\widehat{s})\right)^2}{2(1+r^2)(1+ir\theta_1)\sqrt{1+\theta_1^2r^2}C(\widehat{s})W(\psi_1,\psi_2)(\widehat{s})},$$

where $\{\psi_1, \psi_2\}$ *is a fundamental set of solutions to the second-order linear Eq. (19) and* $\widehat{s} = \frac{2(r^3\theta_2+r\theta_2-\theta_1(-2-r^2\theta_1^2+r^4\theta_1^2))}{(-1-\theta_1^2r^2)^{3/2}}$.

In what follows we focus on obtaining the general solution of Eq. (3) by solving the corresponding transformed Eq. (18). Such solution is implicitly defined by

$$I_1(x,u,u_1,u_2) = C_1, \quad I_2(x,u,u_1,u_2) = C_2, \quad F_1(x,u,u_1,u_2) = C_3, \tag{21}$$

where the expressions of I_1, I_2 and F_1 are given in (20) and $C_i \in \mathbb{C}$ for $i = 1, 2, 3$. Since both functions ψ_1 and ψ_2 are evaluated in $s = (2u_1 + 2u_1^2 + u_2(-u + x))u_1^{-3/2}$, it seems impossible to obtain an explicit solution from (21). With the aim of overcoming such difficulty, we focus on obtaining the general solution in parametric form, as in [22]. Thus, we introduce a new parameter t such that $s = s(t)$ is determined as follows

$$s'(t) = \frac{1}{C(s(t))}.$$

It can be checked that if $\psi_1 = \psi_1(s)$ and $\psi_2 = \psi_2(s)$ is a fundamental set of solutions to the second-order linear Eq. (19), then $\phi_1(t) = \psi_1(s(t))$ and $\phi_2(t) = \psi_2(s(t))$ are two linearly independent solutions to the linear equation

$$\phi''(t) + s(t)\phi'(t) + 4\phi(t) = 0, \tag{22}$$

and the implicit solution (21) becomes

$$\frac{2\sqrt{u_1}\phi_1(t)+\phi_1'(t)}{2\sqrt{u_1}\phi_2(t)+\phi_2'(t)} = C_1, \quad \frac{2\sqrt{u_1}x\phi_1(t)+u\phi_1'(t)}{2\sqrt{u_1}x\phi_2(t)+u\phi_2'(t)} = C_2, \quad \frac{(2x\sqrt{u_1}\phi_2(t)+u\phi_2'(t))^2}{2\sqrt{u_1}(u-x)W(\phi_1,\phi_2)(t)} = C_3. \tag{23}$$

From (23), the following parametrized solution to Eq. (18) is obtained:

$$x(t) = \frac{C_3(C_1-C_2)(C_2\phi_2'(t)-\phi_1'(t))}{C_1\phi_2'(t)-\phi_1'(t)}, \quad u(t) = \frac{C_3(C_1-C_2)(C_2\phi_2(t)-\phi_1(t))}{C_1\phi_2(t)-\phi_1(t)}, \tag{24}$$

where ϕ_1 and ϕ_2 are two linearly independent solutions to Eq. (22), $C_i \in \mathbb{C}$ for $i = 1, 2, 3$ and $C_1 \neq C_2$. As a consequence, by expressing the parametrized solution (24) in terms of the original coordinates by means of (16), the following theorem has been proved:

Theorem 3 *The general solution of a third-order ODE admitting the Lie symmetry algebra* $\mathfrak{so}(3,\mathbb{C})$ *spanned by (14) is given in parametric form through*

$$r(t) = \left(\frac{(C_2\phi_2(t) - \phi_1(t))(C_1\phi_2'(t) - \phi_1'(t))}{(\phi_1(t) - C_1\phi_2(t))(C_2\phi_2'(t) - \phi_1'(t))} \right)^{1/2},$$

$$\theta(t) = \frac{i}{2} \ln \left(\frac{C_3^2(C_1 - C_2)^2(C_2\phi_2(t) - \phi_1(t))(C_2\phi_2'(t) - \phi_1'(t))}{(\phi_1(t) - C_1\phi_2(t))(C_1\phi_2'(t) - \phi_1'(t))} \right), \tag{25}$$

where $C_i \in \mathbb{C}$ for $i = 1, 2, 3$, $C_1 \neq C_2$, and ϕ_1 and ϕ_2 are two linearly independent solutions to Eq. (22).

5 Examples

5.1 *Example I*

As a first example, we consider the following third-order equation:

$$\frac{(r^3\theta_1^5 - r^3\theta_3\theta_1^2 + 3r^3\theta_1\theta_2^2 + 3\theta_1^2 r^2\theta_2 + 4\theta_1^3 r - r\theta_3 - 3\theta_2)(1 + r^2)^2}{(\theta_1^2 r^2 + 1)^3} = 1. \tag{26}$$

It can be checked that the Lie symmetry algebra of Eq. (26) is three-dimensional and corresponds to the nonsolvable algebra $\mathfrak{so}(3, \mathbb{C})$ spanned by the vector fields given in (14). By means of the local change of coordinates (16), Eq. (26) is mapped into:

$$\frac{(x - u)^2(3u_2^2 - 2u_1u_3)}{u_1^3} = -4i, \tag{27}$$

which corresponds with the SL(2)-invariant third-order Eq. (18) for the particular case of

$$C(s) = \frac{i}{4}.$$

Once the function $C = C(s)$ has been identified, we can apply Theorem 2 to obtain the following first integrals to Eq. (26):

$$I_1 = \frac{ir\sqrt{1 + r^2\theta_1^2}\,\psi_1(\widehat{s}) + 2(i\theta_1 r + 1)\psi_1'(\widehat{s})}{ir\sqrt{1 + r^2\theta_1^2}\,\psi_2(\widehat{s}) + 2(i\theta_1 r + 1)\psi_2'(\widehat{s})},$$

$$I_2 = \frac{i\sqrt{1 + r^2\theta_1^2}\,\psi_1(\widehat{s}) - 2(r + ir^2\theta_1)\psi_1'(\widehat{s})}{i\sqrt{1 + r^2\theta_1^2}\,\psi_2(\widehat{s}) - 2(r + ir^2\theta_1)\psi_2'(\widehat{s})},$$

$$F_1 = \frac{e^{-i\theta}\left(i\sqrt{1+r^2\theta_1^2}\psi_2(\widehat{s}) - 2(r+ir^2\theta_1)\psi_2'(\widehat{s})\right)^2}{2i(1+r^2)(1+ir\theta_1)\sqrt{1+\theta_1^2 r^2}W(\psi_1,\psi_2)(\widehat{s})},$$

where $\{\psi_1, \psi_2\}$ is a fundamental set of solutions to the second-order linear equation

$$\psi''(s) - \frac{s}{4i}\psi'(s) - \frac{1}{4}\psi(s) = 0$$

and

$$\widehat{s} = \frac{2\left(r^3\theta_2 + r\theta_2 - \theta_1(-2 - r^2\theta_1^2 + r^4\theta_1^2)\right)}{(-1-\theta_1^2 r^2)^{3/2}}.$$

Now we proceed to compute the general solution to the original Eq. (26) in parametric form. With this aim, we introduce a new parameter t such that $s = s(t)$ is determined as follows:

$$s'(t) = \frac{1}{C(s(t))} = -4i,$$

which yields

$$s(t) = -4it.$$

As a consequence, the general solution to Eq. (26) can be given in parametric form by means of two linearly independent solutions to the second-order linear equation

$$\phi''(t) - 4it\phi'(t) + 4\phi(t) = 0. \tag{28}$$

Let $M(a; b; z)$ and $U(a; b; z)$ denote the Kummer functions of first and second kind, respectively, with parameters values $a = \frac{1}{2} + \frac{1}{2}i$ and $b = \frac{3}{2}$. It can be checked that two linearly independent solutions to Eq. (28) become

$$\phi_1(t) = t\,M(a; b; 2it^2) \quad \text{and} \quad \phi_2(t) = t\,U(a; b; 2it^2). \tag{29}$$

Therefore, the general solution to Eq. (26) is given in parametric from through (25), where the functions ϕ_1 and ϕ_2 are given in (29).

5.2 *Example II*

Let us consider the third-order equation

$$\frac{(\theta_1^5 r^3 - \theta_3 r^3\theta_1^2 + 3r^3\theta_1\theta_2^2 + 3\theta_1^2 r^2\theta_2 - \theta_3 r + 4\theta_1^3 r - 3\theta_2)(r^2+1)^2}{(\theta_1^2 r^2 + 1)^{3/2}(-\theta_2 r - \theta_2 r^3 - 2\theta_1 - \theta_1^3 r^2 + \theta_1^3 r^4)} = 1. \tag{30}$$

It can be checked that its Lie symmetry algebra is three-dimensional and spanned by the vector fields given in (14). By means of the local change of coordinates (16) we obtain the following transformed equation:

$$\frac{(x-u)^2(3u_2^2-2u_1u_3)}{2(2u_1^2+2u_1+xu_2-uu_2)u_1^{3/2}}=1,$$

which corresponds with the SL(2)-invariant third-order ODE (18) for the particular case of

$$C(s)=\frac{-1}{2s},\quad s=(2u_1+2u_1^2+u_2(-u+x))u_1^{-3/2}.$$

In this example we directly computed the general parametric solution, although a complete set of first integrals can be deduced by using Theorem 2 in the same way that in the previous example. The condition

$$s'(t)=\frac{1}{C(s(t))}=-2s(t)$$

yields

$$s(t)=e^{-2t}.$$

By Theorem 3, the general solution to Eq. (30) is given in parametric form through (25), where ϕ_1 and ϕ_2 are two linearly independent solutions to the second-order linear equation

$$\phi''(t)+e^{-2t}\phi'(t)+4\phi(t)=0. \tag{31}$$

If I_{ν_1} and K_{ν_2} stand for the modified Bessel functions of the first and second kinds, respectively, with parameters $\nu_1=-\frac{1}{2}+i$ and $\nu_2=\frac{1}{2}-i$, then it can be checked that two linearly independent solutions to Eq. (31) become:

$$\phi_1(t)=\exp\left\{-\frac{1}{4}e^{-2t}-t\right\}\left(I_{\nu_1}\left(-\frac{1}{4}e^{-2t}\right)+I_{-\overline{\nu_1}}\left(-\frac{1}{4}e^{-2t}\right)\right),$$

$$\phi_2(t)=\exp\left\{-\frac{1}{4}e^{-2t}-t\right\}\left(K_{\nu_2}\left(-\frac{1}{4}e^{-2t}\right)-K_{\overline{\nu_2}}\left(-\frac{1}{4}e^{-2t}\right)\right).$$

These functions ϕ_1 and ϕ_1 permit to obtain the exact general solution to Eq. (30) in parametric form through (25).

6 Concluding Remarks

Solvable structures have been explicitly constructed for third-order ODEs with Lie symmetry algebra isomorphic to $\mathfrak{so}(3, \mathbb{C})$. For this type of equations, the computation of such solvable structures allows to obtain a complete set of first integrals expressed in terms of solutions to a second-order linear equation. The general solution has been also obtained in parametric form and in terms of the solutions to a related second-order linear ODE. The presented results complement the study on the integrability of third-order ODEs admitting non-solvable symmetry algebra of dimension 3 initiated in [20–22] for the case of $\mathfrak{sl}(2, \mathbb{R})$.

Finally, our method has been successfully applied to two examples corresponding with two third-order equations whose Lie symmetry algebra is three-dimensional and isomorphic to $\mathfrak{so}(3, \mathbb{C})$.

Acknowledgements The authors acknowledge the financial support from the University of Cádiz by means of the project PR2017-090 and from the *Junta de Andalucía* research group FQM-377.

A. Ruiz acknowledges the financial support from the Ministry of Education, Culture and Sport of Spain (FPU grant FPU15/02872).

References

1. Abraham-Shrauner, B.: Hidden symmetries and nonlocal group generators for ordinary differential equations IMA. J. Appl. Math. **56**(3), 235–252 (1996)
2. Adam, A.A., Mahomed, F.M.: Integration of ordinary differential equations via nonlocal symmetries. Nonlinear Dyn. **30**(2), 267–275 (2002)
3. Barco, M.A., Prince, G.E.: Solvable symmetry structures in differential form applications. Acta Appl. Math. **66**(1), 89–121 (2001)
4. Basarab-Horwath, P.: Integrability by quadratures for systems of involutive vector fields. Ukr. Math. J. **43**(10), 1236–42 (1991)
5. Catalano Ferraioli, D., Morando, P.: Applications of solvable structures to the nonlocal symmetry-reduction of ODEs. Nonlinear Math. Phys. **16**(suppl.1), 27–42 (2009)
6. Cicogna, G., Gaeta, G., Morando, P.: On the relation between standard and μ-symmetries for PDEs. J. Phys. A Math. Gen. **37**, 9467–9486 (2004)
7. Clarkson, P., Olver, P.J.: Symmetry and the Chazy equation. J. Differ. Equ. **124**(1), 225–46 (1996)
8. Gaeta, G., Morando, P.: On the geometry of lambda-symmetries and PDEs reduction. J. Phys. A Math. Gen. **37**, 6955–6975 (2004)
9. Gusyatnikova, V.N., Yumaguzhin, V.A.: Contact transformations and local reducibility of ODE to the form $y''' = 0$. Acta Appl. Math. **56**(2–3), 155–179 (1999)
10. Hydon, P.E.: Symmetry Methods for Differential Equations: A Beginner's Guide. Cambridge Texts in Applied Mathematics. Cambridge University Press, Cambridge (2000)
11. Ibragimov, N.H.: A Practical Course in Differential Equations and Mathematical Modelling: Classical and New Methods. Nonlinear Mathematical Models Symmetry and Invariance Principles. ALGA Publications (2004)
12. Ibragimov, N.H., Nucci, M.C.: Integration of third order ordinary differential equations by Lie's method: equations admitting three-dimensional Lie algebras. Lie Groups Appl. **1**(2), 49–64 (1994)

13. Kamran, N., Olver, P.J., González-López, A.: Lie algebras of the vector fields in the real plane. Proc. Lond. Math. Soc. **64**, 339–68 (1992)
14. Muriel, C., Romero, J.L.: New methods of reduction for ordinary differential equations. IMA J. Appl. Math. **66**(2), 111–125 (2001)
15. Muriel, C., Romero, J.L.: $\mathcal{C}^\infty$–symmetries and non-solvable symmetry algebras. IMA J. Appl. Math. **66**, 477–498 (2001)
16. Muriel, C., Romero, J.L.: Integrability of equations admitting the nonsolvable symmetry algebra $\mathfrak{so}(3, \mathbb{R})$. Stud. Appl. Math. **109**, 337–352 (2002)
17. Olver, P.J.: Equivalence, Invariants, and Symmetry. Cambridge University Press, Cambridge (1995)
18. Olver, P.J.: Applications of Lie Groups to Differential Equations. Graduate Texts in Mathematics. Springer, Berlin (2000)
19. Prince, G., Sherring, J.: Geometric aspects of reduction of order. Trans. Am. Math. Soc. **334**, 433–453 (1992)
20. Ruiz, A., Muriel, C.: Solvable structures associated to the non-solvable symmetry algebra $\mathfrak{sl}(2, \mathbb{R})$. SIGMA **077**, 18 pp. (2016)
21. Ruiz, A., Muriel, C.: Applications of $\mathcal{C}^\infty$–symmetries in the construction of solvable structures. Trends in Differential Equations and Applications, SEMA SEMAI Springer Series, vol. 8, pp. 387–403 (2016)
22. Ruiz, A., Muriel, C.: First integrals and parametric solutions of third-order ODEs with Lie symmetry algebra isomorphic to $\mathfrak{sl}(2, \mathbb{R})$. J. Phys. A Math. Theor. **50**, 205201 (2017)
23. Stephani, H.: Differential Equations: Their Solution Using Symmetries. Cambridge University Press, Cambridge (1989)

Classification of Scalar Fourth Order Ordinary Differential Equations Linearizable via Generalized Lie–Bäcklund Transformations

Hina M. Dutt and Asghar Qadir

Abstract Lie had shown that there is a unique class of scalar second order ordinary differential equations (ODEs) that can be converted to linear form by point transformations. Mahomed and Leach had shown that for higher order (than 2) scalar ODEs there are always three classes. Separately, Chern had linearized two classes of third order ODEs by using contact transformations. We provided an (inclusive) classification for third order ODEs by using a generalization of contact transformations. Here we extend that work to the fourth order using a generalization of the Lie–Backlund transformation and demonstrate that there are (at least) four classes of fourth order linearizable ODEs.

Keywords Generalized Lie–Bäcklund transformations · Linearization · System of two second order ODEs

1 Introduction

One of the methods for solving nonlinear Differential Equations (DEs) is to transform them to linear form. This procedure, which is called linearization, is a special case of the equivalence problem. Two DEs are said to be equivalent if there exists an invertible transformation which maps one equation into another. The transformation can be a point transformation that involves the change of independent and dependent variables or it can be contact that also depends on the 1st derivative of the dependent variable. The equivalence property puts all DEs into classes of equivalent equations. For 1st order ODEs, the problem is trivial as all 1st order scalar ODEs are equivalent

H. M. Dutt (✉)
Department of Basic Sciences, School of Electrical Engineering and Computer Science, National University of Sciences and Technology, Campus H-12, Islamabad 44000, Pakistan
e-mail: hina.dutt@seecs.edu.pk

A. Qadir
Physics Department School of Natural Sciences, National University of Sciences and Technology, Campus H-12, Islamabad 44000, Pakistan
e-mail: asgharqadir46@gmail.com

V. G. Kac et al. (eds.), *Symmetries, Differential Equations and Applications*, Springer Proceedings in Mathematics & Statistics 266,
https://doi.org/10.1007/978-3-030-01376-9_4

to each other [15]. This is not so for 2nd order scalar ODEs. However, all linear 2nd order scalar ODEs are equivalent and have 8 Lie point symmetry generators. Lie [11–14] showed that a nonlinear 2nd order scalar ODE is transformable to a linear one, i.e. *linearizable*, if it has exactly 8 Lie point symmetries. He proved that a linearizable 2nd order scalar ODE is semilinear and is at most cubic in the 1st derivative with the coefficients of the ODE satisfying four constraints. These constraints, which involve two auxiliary functions, were reduced to two (2nd order constraints) by Tressè [21].

For linearizable nth, $(n \geq 3)$ order scalar ODEs, Mahomed and Leach [17] proved that there are three equivalence classes with $n+1$, $n+2$ and $n+4$ Lie point symmetries. According to Laguerre [9, 10], the canonical form of linear 3rd order scalar ODEs is $y''' + k(x)y = 0$. Chern [1, 2] and Grebot [4, 5] used contact and restricted point transformations respectively, to reduce 3rd order scalar ODEs to the above form with constant k. The general 3rd order linearizable ODE was dealt with by Neut and Petitot [18], and independently by Ibragimov and Meleshko (IM) [6, 7] who used the Cartan approach and original Lie procedure of point transformations respectively, to obtain the desired linearazibility criteria for it. The linearization problem of 4th order scalar ODEs became more complicated and was studied by Ibragimov, Meleshko and Suksern (IMS) [8, 20]. They used point and contact transformations to linearize 4th order scalar ODEs.

As mentioned above, Mahomed and Leach obtained three equivalence classes of nth order scalar ODEs linearizable via point transformations and IMS provided the necessary form of equations linearizable via contact transformations. But, there was no attempt made to obtain classes of equations that are linearizable via higher order derivative transformations. Recently, a new type of transformation, which was called a generalized contact transformation, was introduced by Dutt and Qadir [3]. It was used to obtain equivalence classes of generalized contact symmetries of higher order linearizable scalar ODEs. The equivalence classes of 3rd order ODEs were obtained and found to be four; namely with 5, 6, 7 and 15 generators [3]. Here, we define a new class of transformations which we call generalized Lie–Bäcklund transformations. We reduce scalar ODEs of order n to systems of ODEs of order $\leq n-2$. The point symmetries of the system correspond to the generalized Lie–Bäcklund symmetries of the scalar ODE. This separates all linearizable ODEs in equivalence classes on the basis of the number of generators of higher order derivative transformations. We obtain the canonical form of scalar 4th order ODEs linearizable via these transformations and perform the group classification of these equations.

The outline of the paper is as follows. In the subsequent section, we obtain the canonical form of the 4th order scalar ODEs linearizable via our generalization of the Lie–Bäcklund transformations and perform the group classification of these equations. Concluding remarks are provided in the last section.

2 Generalized Lie–Bäcklund Transformations

A transformation of the form

$$\begin{aligned} t &= \varphi(x, y, y', y'', \ldots, y_i^{(p)}), \\ u &= \psi(x, y, y', y'', \ldots, y^{(p)}), \\ u_i &= \psi_i(x, y, y', y'', \ldots, y^{(p)}), \quad i = 1, 2, \ldots p \end{aligned}$$

is called a *Lie–Bäcklund transformation* of order p if it satisfies the tangency conditions up to order p:

$$u_i = \frac{du_{i-1}}{dt}, \quad i = 1, 2, \ldots p. \tag{1}$$

Lie–Bäcklund transformations depend on independent and dependent variables and derivatives of the dependent variable up to some finite order.

Consider an nth order scalar ODE with ($n \geq 4$)

$$y^{(n)} = f(x, y, y', y'', \ldots, y^{(n-1)}). \tag{2}$$

We define p to be

$$\begin{aligned} 1 < p &\leq \frac{n}{2}, \ if \ n \ is \ even \\ 1 < p &\leq \frac{n-1}{2}, if \ n \ is \ odd. \end{aligned}$$

We replace $y^{(p)} = z$ in (2) to form the following system of two ODEs

$$\begin{aligned} y^{(p)} &= z, \\ z^{(p)} &= f(x, y, z; z'z'', \ldots, z^{(p-1)}), \quad if \ n \ is \ even, \end{aligned}$$

and

$$\begin{aligned} y^{(p+1)} &= z', \\ z^{(p+1)} &= f(x, y, z; z'z'', \ldots, z^{(p)}), \quad if \ n \ is \ odd. \end{aligned}$$

A point transformation

$$t = \varphi(x, y, z), \quad u = \psi(x, y, z), \quad v = \omega(x, y, z), \tag{3}$$

for the above system corresponds to a generalized *Lie–Bäcklund transformation* of order p for the scalar ODE (2) with $y^{(p)} = z$. These transformations depend on the independent, dependent variables and the pth order derivative of the dependent variable but do not require the tangency conditions (1) to hold.

2.1 Group Classification

Consider the general form of a linear, scalar, 4th order ODE

$$y^{(4)} = \pi(x) + \gamma(x)y + \rho(x)y' + \lambda(x)y'' + \varrho(x)y'''. \tag{4}$$

Taking $y'' = z$ will convert the above equation to a system of two 2nd order ODEs

$$\begin{aligned} y'' &= z, \\ z'' &= \pi(x) + \gamma(x)y + \rho(x)y' + \lambda(x)z + \varrho(x)z'. \end{aligned} \tag{5}$$

Any system of second order non-homogeneous ODEs can be mapped invertibly to one of the two canonical forms [22], given by

$$\begin{aligned} y'' &= g_1(x)y + g_2(x)z, \\ z'' &= g_3(x)y + g_4(x)z, \end{aligned} \tag{6}$$

or

$$\begin{aligned} y'' &= k_1(x)y' + k_2(x)z', \\ z'' &= k_3(x)y' + k_4(x)z'. \end{aligned} \tag{7}$$

Since (5) has a nonzero coefficient of z (i.e. 1), so it can only be identified with (6). This makes all coefficient functions in (5) zero except $\gamma(x)$ and $\lambda(x)$. Hence, we have the following reduced system of linear second order ODEs

$$\begin{aligned} y'' &= z, \\ z'' &= \gamma(x)y + \lambda(x)z. \end{aligned} \tag{8}$$

This is the reduced form of the system of ODEs that is obtained from a linear scalar 4th order ODE. The transformation which is used for the conversion of (5)–(8) is provided in [22] [eq. (5)].

To obtain equivalence classes of generalized Lie–Bäcklund symmetries for the scalar Eq. (4), we perform the point symmetry group classification for the system (8). For this purpose, suppose $\mathbf{X}^{(2)}$, given by

$$\begin{aligned} \mathbf{X}^{(2)} = \xi(x,y,z)\frac{\partial}{\partial x} + \eta_1(x,y,z)\frac{\partial}{\partial y} + \eta_2(x,y,z)\frac{\partial}{\partial z} + \eta_1^{(1)}(x,y,z)\frac{\partial}{\partial y'} \\ + \eta_2^{(1)}(x,y,z)\frac{\partial}{\partial z'} + \eta_1^{(2)}(x,y,z)\frac{\partial}{\partial y''} + \eta_2^{(2)}(x,y,z)\frac{\partial}{\partial z''}, \end{aligned} \tag{9}$$

be the symmetry generator of the system (8). The symmetry conditions give us the following set of determining PDEs

$$\xi_{,yy} = \xi_{,yz} = \xi_{,zz} = 0, \quad \eta_{1,zz} = \eta_{2,yy} = 0, \quad \eta_{1,yy} - 2\xi_{,xy} = 0, \tag{10}$$

$$\eta_{2,zz} - 2\xi_{,xz} = 0, \quad \eta_{1,yz} - \xi_{,xz} = 0, \quad \eta_{2,yz} - 2\xi_{,xy} = 0, \quad \eta_{1,xz} - z\xi_{,z} = 0, \tag{11}$$

$$2\eta_{1,xy} - \xi_{,xx} - 3z\xi_{,y} - \gamma y\xi_{,z} - \lambda z\xi_{,z} = 0, \quad \eta_{2,xy} - \lambda y\xi_{,y} - \gamma z\xi_{,y} = 0, \tag{12}$$

$$2\eta_{2,xz} - \xi_{,xx} - z\xi_{,y} - 3\gamma y\xi_{,z} - 3\lambda z\xi_{,z} = 0, \tag{13}$$

$$-\eta_2 + \eta_{1,xx} + z\eta_{1,y} - 2z\xi_{,x} + \gamma y\eta_{1,z} + \lambda z\eta_{1,z} = 0, \tag{14}$$

$$\xi y\gamma_{,x} + \xi z\lambda_{,x} + \eta_1\gamma + \eta_2\lambda - \eta_{2,xx} - z\eta_{2,y} - \gamma y\eta_{2,z} + 2\gamma y\xi_{,x} - \lambda z\eta_{2,z} + 2\lambda z\xi_{,x} = 0. \tag{15}$$

The above system has the following set of solutions

$$\xi = a_1(x), \quad \eta_1 = \left(\frac{a_{1,x}}{2} + c_3\right)y + c_1 z + v(x), \quad \eta_2 = c_2 y + \left(\frac{a_{1,x}}{2} + c_4\right)z + w(x), \tag{16}$$

where (v, w) solves (8), c_i, $(i = 1, 2, 3, 4)$ are arbitrary constants and $a_1(x)$ satisfies

$$a_{1,xxx} + 2c_1\gamma - 2c_2 = 0, \tag{17}$$

$$2a_{1,x} - c_1\lambda - c_3 + c_4 = 0, \tag{18}$$

$$2\gamma a_{1,x} + a_1\gamma_{,x} + (c_3 - c_4)\gamma + c_2\lambda = 0, \tag{19}$$

$$a_{1,xxx} - 4\lambda a_{1,x} - 2\lambda_{,x}a_1 - c_1\gamma + c_2 = 0. \tag{20}$$

We now consider different cases for $\gamma(x)$ to be zero, nonzero constant and an arbitrary function of x.

Case I $\gamma = 0$

With the substitution $\gamma = 0$ in (19), it becomes

$$c_2\lambda = 0, \tag{21}$$

which prompts the consideration of the following cases.

Case I.1 $\gamma = 0, \lambda = 0$

From (18), we get $a_1 = (c_3 - c_4)\frac{x}{2} + c_5$, so that we have $\xi = (c_3 - c_4)\frac{x}{2} + c_5$. Therefore, in this case, we get the following 8−dimensional Lie algebra

$$\mathbf{X}_1 = \frac{\partial}{\partial x}, \quad \mathbf{X}_2 = \frac{\partial}{\partial y}, \quad \mathbf{X}_3 = x\frac{\partial}{\partial y}, \quad \mathbf{X}_4 = y\frac{\partial}{\partial y} + z\frac{\partial}{\partial z}, \tag{22}$$

$$\mathbf{X}_5 = z\frac{\partial}{\partial y}, \quad \mathbf{X}_6 = x\frac{\partial}{\partial x} + 2y\frac{\partial}{\partial y}, \quad \mathbf{X}_7 = \frac{1}{6}x^3\frac{\partial}{\partial y} + x\frac{\partial}{\partial z}, \quad \mathbf{X}_8 = \frac{1}{2}x^2\frac{\partial}{\partial y} + \frac{\partial}{\partial z}.$$

Case I.2 $\gamma = 0, \lambda = \lambda_0 \neq 0$

From (21) we get $c_2 = 0$. From (20) we have $a_{1,x} = 0$ which implies that $a_1 = c_5$. Hence we get a 7−dimensional Lie algebra. The first four operators are given by (22) and the other three are

$$\mathbf{X}_5 = z\frac{\partial}{\partial y} + \lambda_0 z\frac{\partial}{\partial z}, \quad \mathbf{X}_6 = e^{\sqrt{\lambda_0}x}\frac{\partial}{\partial y} + \lambda_0 e^{\sqrt{\lambda_0}x}\frac{\partial}{\partial z}, \quad \mathbf{X}_7 = e^{-\sqrt{\lambda_0}x}\frac{\partial}{\partial y} + \lambda_0 e^{-\sqrt{\lambda_0}x}\frac{\partial}{\partial z}.$$

Case I.3 $\gamma = 0$, $\lambda = \lambda(x)$
By taking $\lambda(x)$ to be an arbitrary function of x, like e^x or $(x \pm c)^m$, $m \neq -2$, we get the following 5–dimensional Lie algebra

$$\mathbf{Y}_i = v_j \frac{\partial}{\partial y} + w_j \frac{\partial}{\partial z}, \quad i = 1, 2, 3, 4, \tag{23}$$
$$\mathbf{Y}_5 = y \frac{\partial}{\partial y} + z \frac{\partial}{\partial z},$$

where (v_j, w_j) are linearly independent solutions of (8). In the latter example, $\lambda = (x \pm c)^{-2}$ comes in the following special case.
Case I.3.1 $\gamma = 0$, $\lambda(x) = (cx + d)^{-2}$
This case produces a 6–dimensional Lie algebra. The first four operators are given by (23), while the extra two operators are

$$\mathbf{Y}_5 = y \frac{\partial}{\partial y} + z \frac{\partial}{\partial z}, \quad \mathbf{Y}_6 = x \frac{\partial}{\partial x} - 2z \frac{\partial}{\partial z}.$$

Case II $\gamma(x) \neq 0$
Here, we have two subcases: either γ is a constant or non constant function of x. This gives rise to the following sub subcases.

Case II.1 $\gamma = \gamma_0$, $\lambda = 0$
This case produces $a_1 = c_5$ and $c_2 = \gamma c_1$. Hence, we have a 7–dimensional Lie algebra. The first four generators are given by (23) and extending generators are

$$\mathbf{Y}_5 = \frac{\partial}{\partial x}, \quad \mathbf{Y}_6 = y \frac{\partial}{\partial y} + z \frac{\partial}{\partial z}, \quad \mathbf{Y}_7 = z \frac{\partial}{\partial y} + \gamma_0 y \frac{\partial}{\partial z}.$$

Case II.2 $\gamma = \gamma_0$, $\lambda = \lambda_0$
In this case, we have a 7–dimensional Lie algebra with $\mathbf{Y}_1$, $\mathbf{Y}_2$, $\mathbf{Y}_3$ and $\mathbf{Y}_4$ given by (23). The additional three operators are

$$\mathbf{Y}_5 = \frac{\partial}{\partial x}, \quad \mathbf{Y}_6 = y \frac{\partial}{\partial y} + z \frac{\partial}{\partial z}, \quad \mathbf{Y}_7 = z \frac{\partial}{\partial y} + (\gamma_0 y + \lambda_0 z) \frac{\partial}{\partial z}.$$

Case II.3 $\gamma = \gamma(x) \neq 0$, $\lambda = \lambda_0$
Here we get a 5–dimensional Lie algebra with first four operators given by (23). The additional operator is

$$\mathbf{Y}_5 = y \frac{\partial}{\partial y} + z \frac{\partial}{\partial z}.$$

Case II.4 $\gamma = \gamma(x) \neq 0$, $\lambda = \lambda(x)$
This case produces the Lie algebra of *case I.3*.

The system of PDEs (10)–(15) provides us four equivalence classes with 5, 6, 7 and 8 symmetries. These Lie point symmetries of (8) correspond to the generalized Lie–Bäcklund symmetries of order 2 for a scalar 4th order linear ODE. Thus we have the following theorem.

Theorem 1 *A linear scalar 4th order ODE can have one of* 5, 6, 7 *and* 8 *generators of the generalized Lie–Bäcklund transformations of order 2.*

3 Conclusion

In this paper, we addressed the problem of classification of ODEs linearizable via higher order derivative transformations which we called generalized Lie–Bäcklund transformations. We obtained four equivalence classes of generators of these transformations for linear scalar 4th order ODEs. We can carry out the same procedure for higher order ODEs to get the classes of systems of ODEs by reducing them to systems of order two lower and finding the point symmetries of the reduced systems. We can also reduce scalar nth order ODEs to systems of lower order with dimension three or more.

In reducing the equations to systems of 2nd order ODEs, we have the advantage of geometric linearization [19], where we can find the solution of the systems easily by employing the coordinate transformations as the linearizing transformations. Similarly, any system of ODEs of order $n \geq 3$ can be reduced in m steps to a system of second order ODEs to use the power of geometry. In this way, we can relate the higher order symmetries of the scalar ODEs with the point symmetries of reduced systems and can find the equivalence classes of the higher order ODEs. The m steps of reduction of an ODE can be shrunk into one step by defining the second or higher order derivative to be a new dependent variable. In this way, the point symmetries of the reduced system correspond to the generalized Lie Bäcklund symmetries of the corresponding scalar ODE.

This procedure could be carried out to reduce scalar nth order ODEs to systems of lower order with three or more dimensions. As an example, consider the scalar 5th order ODE

$$y^{(5)} = f(x, y; y', y'', y''', y^{(4)}). \tag{24}$$

By defining $y'' = z$ and $z' = u$, we can reduce it to a system of three ODEs of order two

$$y'' = z, \quad z'' = u', \quad u'' = f(x, y, z, u; y', u'), \tag{25}$$

and investigate the point symmetries of the reduced system. We can also reduce the scalar ODE (24) to a system of two third ODEs

$$y''' = z'', \quad z''' = f(x, y, z; y', z', z''), \tag{26}$$

by defining $y'' = z'$ and find the Lie point symmetries of the above system. It would be interesting to find a connection between the equivalence classes of the systems (25) and (26) and relate their point symmetries.

Acknowledgements We are grateful to the Higher Education Commission (HEC) of Pakistan for support under their project no. 3054.

References

1. Chern, S.S.: Sur la géométrie d'une équation différentielle du troisème orde. C.R. Acad. Sci. Paris **204**, 1227–1229 (1937)
2. Chern, S.S.: The geometry of the differential equation $y''' = F(x, y, y', y'')$. Sci. Rep. Nat. Tsing Hua Univ. **4**, 97–111 (1940)
3. Dutt, H.M., Qadir, A.: Classification of scalar third order ordinary diferential equations linearizable via generalized contact transformations. Quaest. Math. **41**(1), 15–26 (2018)
4. Grebot, G.: The linearization of third order ODEs, Preprint (1996)
5. Grebot, G.: The characterization of third order ordinary differential equations admitting a transitive fibre preserving point symmetry group. J. Math. Anal. Appl. **206**, 364–388 (1997)
6. Ibragimov, N.H., Meleshko, S.V.: Linearization of third order ordinary differential equations by point transformations. Arch. ALGA **1**, 71–93 (2004)
7. Ibragimov, N.H., Meleshko, S.V.: Linearization of third order ordinary differential equations by point and contact transformations. J. Math. Anal. Appl. **308**, 266–289 (2005)
8. Ibragimov, N.H., Meleshko, S.V., Suksern, S.: Linearization of fourth order ordinary differential equations by point transformations. J. Phys. Math. Theor. **41**(235206), 1–19 (2008)
9. Laguerre, E.: Sur les équations différentielles linéaires du troisième ordre. Comptes Rendus **88**, 116–119 (1879)
10. Laguerre, E.: Sur quelques invariants des èquations diffèrentielles. Comptes Rendus **88**, 224–227 (1879)
11. Lie, S.: Klassifikation und integration von gewöhnlichen differentialgleichungen zwischen x, y, die eine gruppe von transformationen gestatten I, II, II and IV. Archiv for Mathematik, and 9 187, 249, 371, 431 (1883). [Gesammelte Abhandlungen, V, 240, 282, 362 and 432]
12. Lie, S.: Theorie der Transformationsgruppen I, II and III. Teubner, Leipzig (1888). [Reprinted by Chelsea Publishing Company, New York (1970)]
13. Lie, S.: Differential Equations. Chelsea, New York (1967)
14. Lie, S.: Lectures on differential equations with known infinitesimal transformations. Teubner, Leipzig. [In German, Lie's lectures by G. Sheffers (1891)]
15. Mahomed, F.M.: Symmetry group classification of ordinary differential equations: survey of some results. Math. Methods Appl. Sci. **30**, 1995–2002 (2007)
16. Mahomed, F.M., Leach, P.G.L.: Lie algebras associated with second order ordinary differential equations. J. Math. Phys. **30**, 2770–2777 (1989)
17. Mahomed, F.M., Leach, P.G.L.: Symmetry Lie algebra of nth order ordinary differential equations. J. Math. Anal. Appl. **151**, 80–107 (1990)
18. Neut, S., Petitot, M.: La géométrie de l'équation $y''' = f(x, y, y', y'')$. C.R. Acad. Sci. Paris I **335**, 515–518 (2002)
19. Qadir, A.: Linearization of ordinary differential equations by using geometry. SIGMA **3**, 1–7 (2007)
20. Suksern, S., Ibragimov, N.H., Meleshko, S.V.: Criteria for the fourth order ordinary differential equations to be linearizable by contact transformations. Commun. Nonlinear Sci. Numer. Simul. **14**, 2619–2628 (2009)
21. Tressé, A.: Sur les Invariants differentiels des groupes continus de transformations. Acta Math. **18**, 1–88 (1894)
22. Wafo, Soh C., Mahomed, F.M.: Linearization criteria for a system of second order ordinary differential equations. Int. J. Non-Linear Mech. **36**, 671–677 (2001)

On the Symmetries of a Liénard Type Nonlinear Oscillator Equation

R. Mohanasubha, V. K. Chandrasekar, M. Senthilvelan and M. Lakshmanan

Abstract In the contemporary nonlinear dynamics literature, the nonlinear oscillator equation $\ddot{x} + kx\dot{x} + \frac{k^2}{9}x^3 + \tilde{\lambda}x = 0$ is being analyzed in various contexts both classically and quantum mechanically. Classically this nonlinear oscillator equation has been shown to admit three different types of dynamics depending upon the sign and magnitude of the parameter $\tilde{\lambda}$, namely (i) $\tilde{\lambda} = 0$, (ii) $\tilde{\lambda} > 0$ and (iii) $\tilde{\lambda} < 0$. By considering its importance, in this paper, we present the symmetries of its Lagrangian and underlying equation of motion for all the three cases. In particular, we present Lie point symmetries, λ-symmetries, Noether symmetries and telescopic symmetries of this equation. The utility of the symmetries for all the three cases is demonstrated explicitly.

Keywords Nonlinear oscillators · Lie point symmetries · λ-symmetries · Noether symmetries · Telescopic vector fields

1 Introduction

During the past ten years or so considerable interest has been shown on investigating various properties associated with the Liénard type nonlinear oscillator equation,

R. Mohanasubha
Department of Physics, Anna University, Chennai 600025, India
e-mail: subhajeevi@gmail.com

V. K. Chandrasekar
School of Electrical and Electronics Engineering, Centre for Nonlinear Science and Engineering, SASTRA University, Thanjavur 613401, India
e-mail: chandru25nld@gmail.com

M. Senthilvelan · M. Lakshmanan (✉)
Centre for Nonlinear Dynamics, Bharathidasan University, Tiruchirappalli 620024, India
e-mail: lakshman.cnld@gmail.com

M. Senthilvelan
e-mail: senv0000@gmail.com

V. G. Kac et al. (eds.), *Symmetries, Differential Equations and Applications*, Springer Proceedings in Mathematics & Statistics 266,
https://doi.org/10.1007/978-3-030-01376-9_5

$$\Delta(t, x, \dot{x}, \ddot{x}) = \ddot{x} + kx\dot{x} + \frac{k^2}{9}x^3 + \tilde{\lambda}x = 0, \tag{1}$$

where overdot denotes differentiation with respect to t, k and $\tilde{\lambda}$ are arbitrary parameters [4, 8, 12, 15, 22, 24, 27]. Equation (1) arises in the study of equilibrium configurations of a spherical gas cloud acting under the mutual attraction of its molecules and subject to the laws of thermodynamics [3]. Even though a more general equation of this form with time dependent coefficients has been studied long ago considerable interest has been shown on this particular equation when two of the present authors along with Bindu and Pandey have identified it as one of the linearizable equations when a Lie symmetry analysis was carried out on the Liénard type equation $\ddot{x} + f(x)\dot{x} + g(x) = 0$, where $f(x)$ and $g(x)$ are arbitrary functions of x [24]. Originally three of the present authors have proved the integrability of system (1) and demonstrated that this equation admits a conservative non-standard Lagrangian and Hamiltonian description [4]. They have also shown that the frequency of oscillations of this system for $\tilde{\lambda} > 0$ does not depend on the amplitude of oscillations thereby showing that the amplitude dependence of frequency is not necessarily a fundamental property of nonlinear dynamical phenomena [4].

The system (1) admits three different dynamics depending upon the sign of the linear term in it. For example, for the choice $\tilde{\lambda} \leq 0$, the system (1) admits front like solution and $\tilde{\lambda} > 0$ displays explicit sinusoidal periodic solution [4].

This model has further been investigated by several authors under different perspectives [1, 5–7, 9, 15, 16, 22]. For example, it has been demonstrated that the model (1) admits (i) integrating factors [1, 5, 7, 16], (ii) Lagrangian multipliers [22], (iii) $\lambda-$ symmetries [16], (iv) Darboux polynomials [15], and (v) alternate Lagrangian [9]. Equation (1) can be transformed (i) to a free particle equation through invertible point transformation, (ii) to a harmonic oscillator equation through Sundman transformation and (iii) to a linear third order ODE, $w''' + \tilde{\lambda}w = 0$, through a generalized transformation [5–7, 15, 16].

In one of our earlier works, we have constructed a nonstandard Lagrangian, [4]

$$L_1 = \frac{27\tilde{\lambda}^3}{2k^2}\left(\frac{1}{k\dot{x} + \frac{k^2}{3}x^2 + 3\tilde{\lambda}}\right) + \frac{3\tilde{\lambda}}{2k}\dot{x} - \frac{9\tilde{\lambda}^2}{2k^2}, \quad \tilde{\lambda} \neq 0 \tag{2}$$

for this equation. For many of our investigations we stick to the Lagrangian (2) and its associated Hamiltonian (see Eq. (3) below) since when $k \to 0$ both the Lagrangian and Hamiltonian reduce to the linear harmonic oscillator Lagrangian and Hamiltonian, respectively, as the equation of motion does. In a recent work, two of the present authors with Chithiika Ruby have also demonstrated the quantum solvability of its Hamiltonian [8]

$$H = \frac{9\tilde{\lambda}^2}{2k^2}\left(2 - 2\left(1 - \frac{2kp}{3\tilde{\lambda}}\right)^{\frac{1}{2}} + \frac{k^2x^2}{9\tilde{\lambda}} - \frac{2kp}{3\tilde{\lambda}} - \frac{2k^3x^2p}{27\tilde{\lambda}^2}\right), \tag{3}$$

where

$$p = \frac{\partial L}{\partial \dot{x}} = -\frac{27\tilde{\lambda}^3}{2k}\left(\frac{1}{(k\dot{x} + \frac{k^2}{3}x^2 + 3\tilde{\lambda})^2}\right) + \frac{3\tilde{\lambda}}{2k}. \tag{4}$$

By observing that the Hamiltonian (3) can also be equivalently considered in the form

$$H(x, p) = \frac{x^2}{2m(p)} + U(p), \quad -\infty < p \leq \frac{3\tilde{\lambda}}{2k}, \tag{5}$$

where

$$m(p) = \frac{1}{\tilde{\lambda}\left(1 - \frac{2k}{3\tilde{\lambda}}p\right)} \quad \text{and} \quad U(p) = \frac{9\tilde{\lambda}^2}{2k^2}\left(\sqrt{1 - \frac{2k}{3\tilde{\lambda}}p} - 1\right) \tag{6}$$

and recognizing that this form coincides with the position dependent mass Hamiltonian with the difference that the variables x and p are interchanged, the authors went on to quantize the position dependent mass Schroedinger equation in momentum space by augmenting with van Roos ordering. The explicit eigenvalues and eigenvectors have been brought out in an elegant manner.

In this paper we present symmetries of various kinds for Eq. (1) and the nonstandard Lagrangian (2). The reason for consolidating this result is that as far as symmetries are concerned some of the earlier studies are incomplete. For example, eventhough Lie point symmetries are known for this equation for all the three parametric regimes the order reduction procedure has not been done so far for this system. In this paper, we intend to complete it. As far as λ-symmetries are concerned eventhough a detailed investigation has been made on the $\tilde{\lambda} = 0$ case, the analysis has not been carried out for the $\tilde{\lambda} \neq 0$ cases. In this paper we carry out the λ-symmetry analysis for the $\tilde{\lambda} \neq 0$ cases and present two independent λ-symmetries and their associated independent integrals. Similarly eventhough Noether symmetries for the nonstandard Lagrangian (2) with $\tilde{\lambda} = 0$ has been reported it has not been analysed for the $\tilde{\lambda} \neq 0$ cases. We present the Noether symmetries for the remaining two important cases, namely (i) $\tilde{\lambda} > 0$ and (ii) $\tilde{\lambda} < 0$ as well. The telescopic vector fields, which are more generalized vector fields that play important role when the Lie point symmetries and λ-symmetries are absent for a given second order nonlinear ordinary differential equation, are also unknown for this equation. We construct the telescopic vector fields also for Eq. (1).

The plan of the paper is as follows. In Sect. 2, we recall Lie point symmetries of the nonlinear oscillator equation (1) and carry out order reduction procedure for this equation. In Sect. 3, we carry-out the λ-symmetry analysis for the nonlinear oscillator Eq. (1). To begin with, we recall the results that are reported for the case $\tilde{\lambda} = 0$. We then extend the analysis for the cases $\tilde{\lambda} \neq 0$ and give a complete picture. In Sect. 4, we recall Noether's theorem and apply this theorem to the model (1) and derive Noether's symmetries for all the three cases, namely (i) $\tilde{\lambda} > 0$, (ii) $\tilde{\lambda} < 0$ and

(iii) $\tilde{\lambda} = 0$. In Sect. 5, we present the telescopic vector fields for all the three cases. We present our conclusions in Sect. 6.

2 Lie Point Symmetries of Eq. (1)

Let the evolution equation (1) be invariant under the one parameter Lie group of infinitesimal transformations [2, 13, 23]

$$\tilde{t} = t + \varepsilon\xi(t, x) + O(\varepsilon^2), \quad \tilde{x} = x + \varepsilon\eta(t, x) + O(\varepsilon^2), \quad \varepsilon \ll 1, \tag{7}$$

where ξ and η represent the symmetries of Eq. (1) and they are functions of the variables t and x. The associated infinitesimal generator can be written as

$$X = \xi(t, x)\frac{\partial}{\partial t} + \eta(t, x)\frac{\partial}{\partial x}. \tag{8}$$

Equation (1) is invariant under the action of (8) iff

$$X^{(2)}(\Delta)|_{\Delta=0} = 0, \tag{9}$$

where

$$X^{(2)} = \xi\frac{\partial}{\partial t} + \eta\frac{\partial}{\partial x} + \eta^{(1)}\frac{\partial}{\partial \dot{x}} + \eta^{(2)}\frac{\partial}{\partial \ddot{x}} \tag{10}$$

and $\eta^{(1)}$ and $\eta^{(2)}$ are first and second prolongations respectively, whose explicit expressions can be found in Refs. [2, 13]. For the sake of completeness, we present symmetries and order reduction procedure for each one of the cases separately.

2.1 *Case 1:* $\tilde{\lambda} = 0$

First let us consider the choice $\tilde{\lambda} = 0$. The invariance condition (9) reads ($\ddot{x} = \phi(t, x, \dot{x})$)

$$\xi\frac{\partial \phi}{\partial t} + \eta\frac{\partial \phi}{\partial x} + \eta^{(1)}\frac{\partial \phi}{\partial \dot{x}} - \eta^{(2)} = 0. \tag{11}$$

Solving the invariance condition (11), one obtains the following symmetry generators, namely [24]

$$X_1 = \frac{\partial}{\partial t}, \; X_2 = t\left(1-\frac{k}{6}xt\right)\frac{\partial}{\partial t} - \frac{k}{3}x^2 t\left(1-\frac{k}{6}xt\right)\frac{\partial}{\partial x}, \; X_3 = x\frac{\partial}{\partial t} - \frac{k}{3}x^3\frac{\partial}{\partial x},$$
$$X_4 = xt\frac{\partial}{\partial t} + x^2\left(1-\frac{k}{3}xt\right)\frac{\partial}{\partial x}, \; X_5 = -\frac{k}{6}xt^2\frac{\partial}{\partial t} + x\left(1-\frac{k}{3}xt+\frac{k^2}{18}x^2t^2\right)\frac{\partial}{\partial x},$$
$$X_6 = t^2\left(1-\frac{k}{6}xt\right)\frac{\partial}{\partial t} + xt\left(1-\frac{k}{2}xt+\frac{k^2}{18}x^2t^2\right)\frac{\partial}{\partial x},$$
$$X_7 = \frac{k}{2}t^2\left(1-\frac{k}{9}xt\right)\frac{\partial}{\partial t} + \left(1-\frac{k^2}{6}t^2x^2+\frac{k^3}{54}t^3x^3\right)\frac{\partial}{\partial x},$$
$$X_8 = -\frac{k}{6}t^3\left(1-\frac{k}{6}xt\right)\frac{\partial}{\partial t} + t\left(1-\frac{k}{2}xt+\frac{k^2}{9}x^2t^2-\frac{k^3}{108}x^3t^3\right)\frac{\partial}{\partial x}. \quad (12)$$

The nonlinear ODE (1) admits maximal symmetry generators and hence it is linearizable [24]. The symmetry generators constitute $sl(3, R)$ symmetry algebra. Besides several applications, the symmetry generators can also be used to reduce the order of the nonlinear ODE (1). In the following, we demonstrate this procedure by considering the vector field X_3 as an example.

Substituting the expression $\xi = x$ and $\eta = \frac{-k}{3}x^3$ in the characteristic equation $\frac{dt}{\xi} = \frac{dx}{\eta} = \frac{d\dot{x}}{\eta^{(1)}}$ and integrating the resultant equation one finds the invariants u and v as $u = t - \frac{3}{kx}$, and $v = \frac{3x}{k} + \frac{x^3}{\dot{x}}$. The second-order invariant can be derived from the relation $w = \frac{dv}{du}$. Evaluating and simplifying the resultant equation, we arrive at $\frac{dv}{du} = \frac{(kx^3+3x\dot{x})^2}{9\dot{x}^2} = \frac{k^2}{9}v^2$. Integrating this first order differential equation we find $v = -\frac{9}{9I_1+k^2u}$, where I_1 is an integration constant. Substituting the expressions u and v in this solution and rewriting the resultant equation for $\dot{x}$, we end up with

$$\dot{x} - \frac{kx(9I_1x + k^2tx - 3k)}{3(9I_1 + k^2t)} = 0. \quad (13)$$

Integrating Eq. (13) we obtain the general solution of the MEE equation in the following form,

$$x(t) = \frac{6(9I_1 + k^2t)}{kt(18I_1 + k^2t) + 6I_2}, \quad (14)$$

where I_2 is the second integration constant. In a similar manner, one can carry out the order reduction procedure for the rest of the vector fields. Since the procedure is repetitive, one can move on to investigate the other two cases.

2.2 Case 2: $\tilde{\lambda} > 0$

In this case the corresponding symmetry generators are [24]

$$
\begin{aligned}
X_1 &= \frac{\partial}{\partial t}, \quad X_2 = -\frac{k}{3\tilde{\lambda}} \cos\sqrt{\tilde{\lambda}}t \frac{\partial}{\partial t} + \left(\cos\sqrt{\tilde{\lambda}}t - \frac{k}{3\sqrt{\tilde{\lambda}}} x \sin\sqrt{\tilde{\lambda}}t\right) \frac{\partial}{\partial x}, \\
X_3 &= \left(\sin 2\sqrt{\tilde{\lambda}}t + \frac{k}{3\sqrt{\tilde{\lambda}}} x \cos 2\sqrt{\tilde{\lambda}}t\right) \frac{\partial}{\partial t} \\
&\quad + x\left(\sqrt{\tilde{\lambda}} \cos 2\sqrt{\tilde{\lambda}}t - \frac{2k}{3} x \sin 2\sqrt{\tilde{\lambda}}t - \frac{k^2}{9\sqrt{\tilde{\lambda}}} x^2 \cos 2\sqrt{\tilde{\lambda}}t\right) \frac{\partial}{\partial x}, \\
X_4 &= \left(\cos 2\sqrt{\tilde{\lambda}}t - \frac{k}{3\sqrt{\tilde{\lambda}}} x \sin 2\sqrt{\tilde{\lambda}}t\right) \frac{\partial}{\partial t} \\
&\quad - x\left(\sqrt{\tilde{\lambda}} \sin 2\sqrt{\tilde{\lambda}}t + \frac{2k}{3} x \cos 2\sqrt{\tilde{\lambda}}t - \frac{k^2}{9\sqrt{\tilde{\lambda}}} x^2 \sin 2\sqrt{\tilde{\lambda}}t\right) \frac{\partial}{\partial x}, \\
X_5 &= x \sin\sqrt{\tilde{\lambda}}t \frac{\partial}{\partial t} + x^2 \left(\sqrt{\tilde{\lambda}} \cos\sqrt{\tilde{\lambda}}t - \frac{k}{3} x \sin\sqrt{\tilde{\lambda}}t\right) \frac{\partial}{\partial x}, \\
X_6 &= x \cos\sqrt{\tilde{\lambda}}t \frac{\partial}{\partial t} - x^2 \left(\sqrt{\tilde{\lambda}} \sin\sqrt{\tilde{\lambda}}t + \frac{k}{3} x \cos\sqrt{\tilde{\lambda}}t\right) \frac{\partial}{\partial x}, \\
X_7 &= x \frac{\partial}{\partial t} - \left(\frac{3\tilde{\lambda}x}{k} + \frac{k}{3} x^3\right) \frac{\partial}{\partial x}, \\
X_8 &= -\frac{k}{3\tilde{\lambda}} \sin\sqrt{\tilde{\lambda}}t \frac{\partial}{\partial t} + \left(\sin\sqrt{\tilde{\lambda}}t + \frac{k}{3\sqrt{\tilde{\lambda}}} x \cos\sqrt{\tilde{\lambda}}t\right) \frac{\partial}{\partial x}.
\end{aligned} \tag{15}
$$

The vector fields (15) can again be shown to form an $sl(3, R)$ algebra.

In the following, we demonstrate the usefulness of the symmetry vector fields by considering the vector field X_7. The associated characteristic equation reads

$$
\frac{dt}{x} = \frac{dx}{-(\frac{kx^3}{3} + \frac{3\tilde{\lambda}x}{k})} = \frac{d\dot{x}}{-\frac{\dot{x}(k^2x^2 + k\dot{x} + 3\tilde{\lambda})}{k}}. \tag{16}
$$

Now integrating the above Eq. (16) we find the invariants u and v to be of the following forms:

$$
u = \frac{\tan^{-1}\left(\frac{kx}{3\sqrt{\tilde{\lambda}}}\right) + \sqrt{\tilde{\lambda}}t}{\sqrt{\tilde{\lambda}}}, \quad v = \frac{x(k^2x^2 + 3k\dot{x} + 9\tilde{\lambda})}{\dot{x}}. \tag{17}
$$

The second-order invariant reads $\frac{dv}{du} = 9\tilde{\lambda} + \frac{v^2}{9}$. Integrating the later we find $v = 9\sqrt{\tilde{\lambda}}\tan(9I_1\sqrt{\tilde{\lambda}} + \sqrt{\tilde{\lambda}}u)$, where I_1 is an integration constant. Substituting the expressions u and v in this solution and rewriting the resultant equation for $\dot{x}$, we end up with

$$\dot{x} + \frac{k^2x^3 + 9\tilde{\lambda}x}{3\left(kx - 3\sqrt{\tilde{\lambda}}\tan\left(\sqrt{\tilde{\lambda}}(9I_1 + t) + \tan^{-1}\left(\frac{kx}{3\sqrt{\tilde{\lambda}}}\right)\right)\right)} = 0. \tag{18}$$

Integrating Eq. (18) we obtain the general solution of (1) with positive values of $\tilde{\lambda}$ in the following form,

$$x(t) = \frac{3\sqrt{\tilde{\lambda}}\sin(\sqrt{\tilde{\lambda}}(9I_1 + t))}{I_2\sqrt{\tilde{\lambda}} - k\cos(\sqrt{\tilde{\lambda}}(9I_1 + t))}, \tag{19}$$

where I_2 is the second integration constant. One may extend the order reduction procedure for the remaining vector fields too in a similar fashion. Now we move on to the third case.

2.3 Case 3: $\tilde{\lambda} < 0$

In this case, we find the equation is invariant under the following forms of symmetry generators: [24]

$$\begin{aligned}
X_1 &= \frac{\partial}{\partial t}, \quad X_2 = e^{2\sqrt{\tilde{\lambda}}t}\left[\left(1 - \frac{k}{3\sqrt{\tilde{\lambda}}}x\right)\frac{\partial}{\partial t} + \left[\left(\frac{k}{9\sqrt{\tilde{\lambda}}}x^3 - \frac{2k}{3}x^2 + \sqrt{\tilde{\lambda}}x\right)\frac{\partial}{\partial x}\right],\right.\\
X_3 &= e^{-2\sqrt{\tilde{\lambda}}t}\left[\left(1 + \frac{k}{3\sqrt{\tilde{\lambda}}}x\right)\frac{\partial}{\partial t} - \left(\frac{k}{9\sqrt{\tilde{\lambda}}}x^3 + \frac{2k}{3}x^2 + \sqrt{\tilde{\lambda}}x\right)\frac{\partial}{\partial x}\right],\\
X_4 &= x\frac{\partial}{\partial t} - \left(\frac{k}{3}x^3 - \frac{3\tilde{\lambda}}{k}x\right)\frac{\partial}{\partial x}, \quad X_5 = e^{\sqrt{\tilde{\lambda}}t}\left[x\frac{\partial}{\partial t} - \left(\frac{k}{3}x^3 - \sqrt{\tilde{\lambda}}x^2\right)\frac{\partial}{\partial x}\right],\\
X_6 &= e^{-\sqrt{\tilde{\lambda}}t}\left[x\frac{\partial}{\partial t} - \left(\frac{k}{3}x^3 + \sqrt{\tilde{\lambda}}x^2\right)\frac{\partial}{\partial x}\right], \quad X_7 = e^{\sqrt{\tilde{\lambda}}t}\left[\frac{\partial}{\partial t} - \left(\sqrt{\tilde{\lambda}}x - \frac{3\tilde{\lambda}}{k}\right)\frac{\partial}{\partial x}\right],\\
X_8 &= e^{-\sqrt{\tilde{\lambda}}t}\left[\frac{\partial}{\partial t} + \left(\sqrt{\tilde{\lambda}}x + \frac{3\tilde{\lambda}}{k}\right)\frac{\partial}{\partial x}\right].
\end{aligned} \tag{20}$$

To obtain the general solution for this case, we consider the vector field X_6. Solving the characteristic equation associated with this vector field

$$\frac{dt}{xe^{-\sqrt{\tilde{\lambda}}t}} = \frac{dx}{-e^{-\sqrt{\tilde{\lambda}}t}\left(\frac{kx^3}{3}+\sqrt{\tilde{\lambda}}x^2\right)}$$
$$= \frac{d\dot{x}}{\frac{1}{3}e^{-\sqrt{\tilde{\lambda}}t}(kx^2(\sqrt{\tilde{\lambda}}x-3\dot{x})-3(-\tilde{\lambda}x^2+\sqrt{\tilde{\lambda}}x\dot{x}+\dot{x}^2))} \tag{21}$$

we obtain the invariants u and v of the form

$$u = \frac{-\log(kx+3\sqrt{\tilde{\lambda}})+\sqrt{\tilde{\lambda}}t+\log(x)}{\sqrt{\tilde{\lambda}}}, \tag{22}$$

$$v = \frac{3(-k\sqrt{\tilde{\lambda}}x^2+2kx\dot{x}-3\tilde{\lambda}x+3\sqrt{\tilde{\lambda}}\dot{x})}{2(kx^4+3\sqrt{\tilde{\lambda}}x^3+3x^2\dot{x})}. \tag{23}$$

The second-order invariant can be found from the relation $w = \frac{dv}{du}$. In this case, we find

$$\frac{dv}{du} = -\frac{(kx+3\sqrt{\tilde{\lambda}})^2(kx^2-3\sqrt{\tilde{\lambda}}x+3\dot{x})}{3x^2(kx^2+3(\sqrt{\tilde{\lambda}}x+\dot{x}))} = -\left(\frac{k^2}{3}+2\sqrt{\tilde{\lambda}}v\right). \tag{24}$$

Integrating Eq. (24), we find $v = e^{-2\sqrt{\tilde{\lambda}}u}I_1 - \frac{k^2}{6\sqrt{\tilde{\lambda}}}$, where I_1 is an integration constant. Substituting the expressions u and v in this solution and rewriting the resultant equation for $\dot{x}$, we end up with

$$\dot{x} + \frac{x(e^{2\sqrt{\tilde{\lambda}}t}(kx-3\sqrt{\tilde{\lambda}})-6I_1(k\sqrt{\tilde{\lambda}}x+3\tilde{\lambda}))}{3(e^{2\sqrt{\tilde{\lambda}}t}-6I_1\sqrt{\tilde{\lambda}})} = 0. \tag{25}$$

Integrating Eq. (25), we obtain the general solution of Eq. (1) with negative $\tilde{\lambda}$ in the following form

$$x(t) = \frac{3(\sqrt{\tilde{\lambda}}e^{2\sqrt{\tilde{\lambda}}t}-6I_1\tilde{\lambda})}{3I_2\sqrt{\tilde{\lambda}}e^{\sqrt{\tilde{\lambda}}t}+6I_1\sqrt{\tilde{\lambda}}k+ke^{2\sqrt{\tilde{\lambda}}t}}, \tag{26}$$

where I_2 is the second integration constant. One may verify that the remaining vector fields can also be used to derive the above general solution of the given Eq. (1).

3 λ-Symmetries

Recently efforts have been made to generalize the classical Lie algorithm and obtain integrals and general solution of nonlinear ODEs, in particular equations which lack Lie point symmetries. One such generalization is the λ-symmetry approach [17].

The method of finding λ-symmetries for a second-order ODE has been discussed in depth by Muriel and Romero [18] and the advantage of finding such symmetries has also been demonstrated by them. They also have developed an algorithm to determine integrating factors and integrals from λ-symmetries for second-order ODEs [19]. The relation among λ-symmetries, Lie point symmetries and local-nonlocal transformations for Liénard I and II-type equations was studied in Ref. [25]. The vector fields associated with λ-symmetries are being denoted as v instead of X just to differentiate λ-symmetries from Lie point symmetry vector fields.

A vector field v is a λ-symmetry of the second-order equation if there exists a function such that

$$v^{[\lambda,(2)]}(\Delta(t,x,\dot{x},\ddot{x})) = 0 \quad when \quad \Delta(t,x,\dot{x},\ddot{x}) = 0, \tag{27}$$

where $v^{[\lambda,(2)]}$ is given by

$$v^{[\lambda,(2)]} = \xi(t,x)\frac{\partial}{\partial t} + \eta^{[\lambda,(0)]}(t,x)\frac{\partial}{\partial x} + \eta^{[\lambda,(1)]}(t,x,\dot{x})\frac{\partial}{\partial \dot{x}} + \eta^{[\lambda,(2)]}(t,x,\dot{x},\ddot{x})\frac{\partial}{\partial \ddot{x}}, \tag{28}$$

with

$$\eta^{[\lambda,(0)]} = \eta(t,x), \tag{29}$$

$$\eta^{[\lambda,(1)]} = (D_t+\lambda)\eta^{[\lambda,(0)]}(t,x) - (D_t+\lambda)(\xi)\dot{x}, \tag{30}$$

$$\eta^{[\lambda,(2)]} = (D_t+\lambda)\eta^{[\lambda,(1)]}(t,x,\dot{x}) - (D_t+\lambda)(\xi)\ddot{x}. \tag{31}$$

In the above prolongation formula if we put $\lambda = 0$, we end up with standard Lie prolongation expressions. Solving the invariance condition (27) we can determine the functions ξ, η and λ for the given equation. We note here that three unknowns ξ, η and λ have to be determined from the invariance condition (27). The procedure is as follows.

Let us suppose that the second-order Eq. (1) has Lie point symmetries. In this case, the λ-function can be determined in a more simple way without solving the invariance condition (27) as follows. If X is a Lie point symmetry of (1) and $Q = \eta - \dot{x}\xi$ is its characteristics, then $v = \frac{\partial}{\partial x}$ is a λ-symmetry of (1) for $\lambda = \frac{D[Q]}{Q}$ [25]. The λ-symmetry satisfies the invariance condition [19]

$$\phi_x + \lambda\phi_{\dot{x}} = D[\lambda] + \lambda^2. \tag{32}$$

Once the λ-symmetry is determined, we can obtain the first integrals in two different ways. In the first way, we can calculate the integral directly from the λ-symmetry using the four step algorithm given below. In the second way, we can find the integrating factor μ from λ-symmetry directly. With the help of integrating factors and λ-symmetries we can obtain the first integrals by integrating the system of Eq. (34) given below. In the following, we enumerate both the procedures.

(A) Method of finding the first integral directly from λ-symmetry [19]

The method of finding the integral directly from λ-symmetry is as follows:

1. Find a first integral $w(t, x, \dot{x})$ of $v^{[\lambda,(1)]}$, that is a particular solution of the equation $w_x + \lambda w_{\dot{x}} = 0$, where the subscript denotes partial derivative with respect to that variable and $v^{[\lambda,(1)]}$ is the first-order λ-prolongation of the vector field v.
2. Evaluate $D[w]$ and express it in terms of (t, w) as $D[w] = F(t, w)$.
3. Find a first integral G of $\partial_t + F(t, w)\partial_w$.
4. Evaluate $I(t, x, \dot{x}) = G(t, w(t, x, \dot{x}))$.

(B) Method of finding integrating factors from λ [19]

If X is a Lie point symmetry of (1) *and* $Q = \eta - \dot{x}\xi$ *is its characteristics, then* $v = \partial_x$ *is a λ-symmetry of* (1) *for* $\lambda = D[Q]/Q$ *and any solution of the first-order linear system*

$$D[\mu] + \left(\phi_{\dot{x}} - \frac{D[Q]}{Q}\right)\mu = 0, \quad \mu_x + \left(\frac{D[Q]}{Q}\mu\right)_{\dot{x}} = 0, \tag{33}$$

is an integrating factor of (1). *Here D represents the total derivative operator and it is given by* $\frac{\partial}{\partial t} + \dot{x}\frac{\partial}{\partial x} + \phi\frac{\partial}{\partial \dot{x}}$.

Solving the system of equations (33) one can get μ. Once the integrating factor μ is known then a first integral I such that $I_{\dot{x}} = \mu$ can be found by solving the system of equations

$$I_t = \mu(\lambda\dot{x} - \phi), \quad I_x = -\lambda\mu, \quad I_{\dot{x}} = \mu. \tag{34}$$

From the first integrals, we can write the general solution of the given equation.

In the following we apply the above method to Eq. (1)

3.1 Case 1: $\tilde{\lambda} = 0$

Bhuvaneshwari et al. had studied the λ-symmetries for Eq. (1) with $\tilde{\lambda} = 0$ [1]. They have found the λ-symmetries from the Lie point symmetries by using the relation $\lambda = \frac{D[Q]}{Q}$, where $Q = \eta - \dot{x}\xi$. For this purpose they considered the Lie point symmetries X_2 and X_4 from Eq. (12). The expressions for Q turns out to be

$$Q_1 = \frac{1}{18}(k^2t^2x^3 - 6ktx^2 + 3kt^2x\dot{x} - 18t\dot{x}), \quad Q_2 = x^2\left(1 - \frac{k}{3}xt\right) - tx\dot{x}. \tag{35}$$

The two λ-functions are of the form

$$\lambda_1 = \frac{\left(1 - \frac{2}{3}ktx - \frac{k}{6}t^2\dot{x} + \frac{k^2}{18}t^2x^2\right)}{t\left(1 - \frac{k}{6}tx\right)}, \ \lambda_2 = \frac{\dot{x}}{x} - \frac{kx}{3}. \tag{36}$$

The associated λ-symmetry is $v = \frac{\partial}{\partial x}$.

3.1.1 First Integrals from λ_1 and λ_2

By following the above discussed procedure, we have found

$$w(t, x, \dot{x}) = \frac{\left(1 - \frac{1}{3}ktx + \frac{k}{6}t^2\dot{x} + \frac{k^2}{18}t^2x^2\right)}{kt^2\left(1 - \frac{k}{6}tx\right)}. \tag{37}$$

In the second step, we obtain determining equation for w as

$$D[w] = ktw^2 - \frac{3w}{t} \tag{38}$$

using λ_1. In the third step, we obtain the function $G(t, w)$ as

$$G(t, w) = \frac{1}{t^3 w} - \frac{k}{t}. \tag{39}$$

In the final step, we found the integral I_1 as

$$I_1 = \frac{k}{6}t + \frac{\left(1 - \frac{k}{6}tx\right)}{\left(t\dot{x} - x + \frac{k}{3}tx^2\right)}. \tag{40}$$

In the same way, we have found the function w for λ_2 as

$$w(x, \dot{x}) = \frac{\dot{x}}{x} + \frac{k}{3}x. \tag{41}$$

In the second step, we get the determining equation as

$$D[w] + w^2 = 0. \tag{42}$$

We get the function G in the third step as

$$G = t - \frac{1}{w}. \tag{43}$$

As the final step we get the integral as

$$I_2 = t - \frac{x}{\dot{x} + \frac{k}{3}x^2}. \tag{44}$$

From the integrals I_1 and I_2, we can write the general solution as

$$x(t) = \frac{t + I_2}{\frac{k}{6}t^2 + \frac{k}{3}tI_2 - I_1 I_2}. \tag{45}$$

3.1.2 Integrating Factors from λ_1 and λ_2

We can also find the integrating factors from λ_1 and λ_2 using the relation (33). Substituting the function λ_1 in Eq. (33) we get

$$\mu_{1x} + \left(\frac{\left(1 - \frac{2}{3}ktx - \frac{k}{6}t^2\dot{x} + \frac{k^2}{18}t^2x^2\right)}{t\left(1 - \frac{k}{6}tx\right)} \mu_1 \right)_{\dot{x}} = 0. \tag{46}$$

The characteristic equation associated with Eq. (46) is given by

$$\frac{dx}{1} = \frac{d\dot{x}}{\frac{\left(1 - \frac{2}{3}ktx - \frac{k}{6}t^2\dot{x} + \frac{k^2}{18}t^2x^2\right)}{t\left(1 - \frac{k}{6}tx\right)}} = \frac{d\mu_1}{-\frac{kt^2}{6\left(t - \frac{1}{6}kt^2x\right)}\mu_1}. \tag{47}$$

Integrating (47) we find the integrals to be of the form

$$C_1 = \frac{\left(1 - \frac{1}{3}ktx + \frac{k}{6}t^2\dot{x} + \frac{k^2}{18}t^2x^2\right)}{kt^2\left(1 - \frac{k}{6}tx\right)}, \quad C_2 = \left(t - \frac{1}{6}kt^2x\right)\mu_1. \tag{48}$$

From the above, we obtain the general solution as

$$\mu_1 = -\frac{C_1\left[t\left(\frac{\left(1 - \frac{1}{3}ktx + \frac{k}{6}t^2\dot{x} + \frac{k^2}{18}t^2x^2\right)}{kt^2\left(1 - \frac{k}{6}tx\right)}\right)\right]}{\left(t - \frac{1}{6}kt^2x\right)}. \tag{49}$$

Choosing the function C_1 appropriately we get

$$\mu_1 = -\frac{k^2t\left(1 - \frac{k}{6}tx\right)}{6\left(1 - \frac{1}{3}ktx + \frac{k}{6}t^2\dot{x} + \frac{k^2}{18}t^2x^2\right)^2}. \tag{50}$$

We find that the expression (50) also satisfies the first equation in (33) as well and thus forms a compatible solution to the system of equation (33).

To determine the integrating factor associated with λ_2 directly we first solve the second equation in (33), that is

$$\mu_{2x} + \left(\frac{\dot{x}}{x} - \frac{k}{3}x\right)\mu_{2\dot{x}} + \frac{1}{x}\mu_2 = 0. \tag{51}$$

The characteristic equation associated with the above equation can be written as

$$\frac{dx}{1} = \frac{d\dot{x}}{\frac{\dot{x}}{x} - \frac{k}{3}x} = \frac{d\mu_2}{\frac{\mu_2}{x}}. \tag{52}$$

Integrating (52) we find the integral as

$$\mu_2 = -\frac{x}{\left(\dot{x} + \frac{k}{3}x^2\right)^2}. \tag{53}$$

We find that the above expression also satisfies the first equation in (33).

3.2 Case 2: $\tilde{\lambda} \neq 0$

In the earlier case where $\tilde{\lambda} = 0$, we fixed the λ-symmetries from the set of Lie point symmetries itself. For the two cases $\tilde{\lambda} > 0$ and $\tilde{\lambda} < 0$ we derive the λ-symmetries by solving the associated invariance condition which has not been considered so far for this equation. To determine the λ-symmetry for Eq. (1), we solve the following determining equation

$$D[\lambda] + \lambda^2 + \lambda k x + k\dot{x} + \frac{k^2}{3}x^2 + \tilde{\lambda} = 0. \tag{54}$$

To obtain a particular solution of Eq. (54), we assume an ansatz

$$\lambda = a_1\dot{x} + a_2, \tag{55}$$

where a_1 and a_2 are functions of x.

Substituting (55) in (54) and solving the resultant equation, we find

$$\lambda_1 = \frac{\dot{x}}{x} - \frac{kx}{3}. \tag{56}$$

Now we use the above said procedure and obtain the first integral. The calculations are given below.

In the first step, we setup the determining equation for $w(t, x, \dot{x})$, that is

$$w_x + \left(\frac{\dot{x}}{x} - \frac{kx}{3}\right) w_{\dot{x}} = 0. \tag{57}$$

A particular solution of (57) is

$$w(t, x, \dot{x}) = \frac{kx}{3} + \frac{\dot{x}}{x}. \tag{58}$$

In the second step, we express $D[w]$ in terms of (t, w) as $D[w] = F(t, w)$. In this case, we find

$$D[w] = -(w^2 + \tilde{\lambda}). \tag{59}$$

In the third step, we fix the function $G(t, w)$ as

$$G(t, w) = \frac{\sqrt{\tilde{\lambda}}t + \tan^{-1}\left(\frac{w}{\sqrt{\tilde{\lambda}}}\right)}{\sqrt{\tilde{\lambda}}}. \tag{60}$$

Now replacing w with the expression (58) we obtain the first integral in the form

$$I(t, x, \dot{x}) = \frac{\tan^{-1}\left(\frac{kx^2+3\dot{x}}{3\sqrt{\tilde{\lambda}}x}\right) + \sqrt{\tilde{\lambda}}t}{\sqrt{\tilde{\lambda}}}. \tag{61}$$

By recalling the formula $\arctan(x) = \frac{1}{2}i[\ln(1 - ix) - ln(1 + ix)]$ and simplifying the resultant equation we obtain the first integral as

$$I_1 = e^{-2\sqrt{-\tilde{\lambda}}t}\frac{\left(\dot{x} + \frac{k}{3}x^2 + x\sqrt{-\tilde{\lambda}}\right)}{\left(\dot{x} + \frac{k}{3}x^2 - x\sqrt{-\tilde{\lambda}}\right)}. \tag{62}$$

To prove the integrability of Eq. (1), we are in need of one more λ-symmetry. To obtain it, we assume a more general ansatz for λ which is of the form

$$\lambda_2 = \frac{a_1(t, x)\dot{x} + a_2(t, x)}{a_3(t, x)\dot{x} + a_4(t, x)}. \tag{63}$$

where a_1, a_2, a_3 and a_4 are arbitrary functions of t and x and to be determined. Substituting the above ansatz in the λ-determining Eq. (54) and solving the resultant equation, we obtain

$$\lambda_2 = \frac{\frac{k\dot{x}}{3} - \left(\frac{kx}{3} + \sqrt{-\tilde{\lambda}}\right)^2}{\frac{kx}{3} + \sqrt{-\tilde{\lambda}}}. \tag{64}$$

We note here that while solving the Eq. (54) with the ansatz (63) we also obtain (56) as another particular solution. We do not mention it here as we have already dealt with it. Following the above said procedure now we find the integral associated with $\tilde{\lambda}_2$. To begin it, we set up the determining equation for $w(t, x, \dot{x})$ as

$$w_x + \frac{\frac{k\dot{x}}{3} - \left(\frac{kx}{3} + \sqrt{-\tilde{\lambda}}\right)^2}{\frac{kx}{3} + \sqrt{-\tilde{\lambda}}} w_{\dot{x}} = 0. \tag{65}$$

A particular solution of the above equation is

$$w(t, x, \dot{x}) = \frac{kx^2 + 3\sqrt{-\tilde{\lambda}}x + 3\dot{x}}{3(kx + 3\sqrt{-\tilde{\lambda}})}. \tag{66}$$

The total derivative of $w(t, x, \dot{x})$ reads

$$D[w] = \sqrt{-\tilde{\lambda}}w - kw^2. \tag{67}$$

In the third step, we determine the function $G(t, w)$ as

$$G(t, w) = -\frac{i \log\left(\frac{e^{i\sqrt{\tilde{\lambda}}t}\left(kw - i\sqrt{\tilde{\lambda}}\right)}{w}\right)}{\sqrt{\tilde{\lambda}}}. \tag{68}$$

Now replacing the variable w by (66) we obtain the integral associated with $\tilde{\lambda}_2$ in the form

$$I(t, x, \dot{x}) = -\frac{i \log\left(\frac{e^{i\sqrt{\tilde{\lambda}}t}(k(3\dot{x}+kx^2)+9\tilde{\lambda})}{kx^2+3\left(\sqrt{-\tilde{\lambda}}x+\dot{x}\right)}\right)}{\sqrt{\tilde{\lambda}}}. \tag{69}$$

After rearranging the integral in more elegant form, we obtain

$$I_2 = -\frac{6}{k}e^{\sqrt{-\tilde{\lambda}}t}\left(\frac{\tilde{\lambda} + \frac{k}{3}\dot{x} + \frac{k^2}{9}x^2}{\dot{x} + \frac{k}{3}x^2 + x\sqrt{-\tilde{\lambda}}}\right). \tag{70}$$

From the integrals I_1 and I_2, we can write the solution of Eq. (1) for $\tilde{\lambda} > 0$ and $\tilde{\lambda} < 0$. First let us consider the case $\tilde{\lambda} > 0$.

3.2.1 Case 2: $\tilde{\lambda} > 0$

For $\tilde{\lambda} > 0$, integrals (62) and (70) are complex. To get the real integrals, we consider the following combinations of the integrals

$$\tilde{I}_1 = \frac{4}{kI_1^2 I_2^2} = \frac{(3\dot{x} + kx^2)^2 + 9\tilde{\lambda}x^2}{(3k\dot{x} + k^2x^2 + 9\tilde{\lambda})^2}, \tag{71}$$

$$\tilde{I}_2 = -\frac{2e^{i\delta}}{k|I_1 I_2|} = e^{i(\sqrt{\tilde{\lambda}}t+\delta)}\left(\frac{3\dot{x} + kx^2 - 3i\sqrt{\tilde{\lambda}}x}{3k\dot{x} + k^2x^2 + 9\tilde{\lambda}}\right), \tag{72}$$

where δ is phase constant. Now the integrals $\tilde{I}_1$ and $|\tilde{I}_2|$ can be considered as two real integrals of Eq. (1) for $\tilde{\lambda} > 0$. The solution for Eq. (1) from the two integrals (62) and (70) can be written as

$$x(t) = \frac{A\sin(\sqrt{\tilde{\lambda}}t + \delta)}{1 - \frac{k}{3\sqrt{\tilde{\lambda}}}A\cos(\sqrt{\tilde{\lambda}}t + \delta)}, \qquad 0 \le A < \frac{3\sqrt{\tilde{\lambda}}}{k}, \tag{73}$$

where $A = 3\sqrt{\tilde{\lambda}\tilde{I}_1}$ and δ is an arbitrary constant.

3.2.2 Case 3: $\tilde{\lambda} < 0$

For $\tilde{\lambda} < 0$, integrals (62) and (70) are real from which we can straightforwardly write the general solution as

$$x(t) = \frac{3\sqrt{|\tilde{\lambda}|(\tilde{I}_1 e^{2\sqrt{|\tilde{\lambda}|}t} - 1)}}{k\tilde{I}_1\tilde{I}_2 e^{\sqrt{|\tilde{\lambda}|}t} + k(1 + \tilde{I}_1 e^{2\sqrt{|\tilde{\lambda}|}t})}, \tag{74}$$

where I_1 and I_2 are constants.

3.2.3 Integrating Factors from λ_1 and λ_2

To find the integrating factors from λ_1 and λ_2, we consider the second equation in (33) and obtain

$$\mu_{1x} + \left(\frac{\dot{x}}{x} - \frac{k}{3}x\right)\mu_{1\dot{x}} + \frac{1}{x}\mu_1 = 0, \tag{75}$$

$$\mu_{2x} + \left(\frac{\frac{k\dot{x}}{3} - \left(\frac{kx}{3} + \sqrt{-\tilde{\lambda}}\right)^2}{\frac{kx}{3} + \sqrt{-\tilde{\lambda}}}\right)\mu_{1\dot{x}} + \frac{k}{3\left(\frac{kx}{3} + \sqrt{-\tilde{\lambda}}\right)}\mu_2 = 0. \tag{76}$$

The characteristic equations associated with the above equations can be written as

$$\frac{dx}{1} = \frac{d\dot{x}}{\frac{\dot{x}}{x} - \frac{k}{3}x} = \frac{d\mu_1}{\frac{\mu_1}{x}}, \tag{77}$$

$$\frac{dx}{1} = \frac{d\dot{x}}{\frac{\frac{k\dot{x}}{3} - \left(\frac{kx}{3} + \sqrt{-\tilde{\lambda}}\right)^2}{\frac{kx}{3} + \sqrt{-\tilde{\lambda}}}} = \frac{d\mu_2}{\frac{3\mu_2\left(\frac{kx}{3} + \sqrt{-\tilde{\lambda}}\right)}{k}}. \tag{78}$$

Solving the above characteristic equations and choosing the constants appropriately, we obtain the solutions of the above equations as

$$\mu_1 = -\frac{18\sqrt{-\tilde{\lambda}}xe^{-2\sqrt{-\tilde{\lambda}}t}}{(kx^2 - 3\sqrt{-\tilde{\lambda}}x + 3\dot{x})^2}, \tag{79}$$

$$\mu_2 = -\frac{18e^{\sqrt{-\tilde{\lambda}}t}(k\sqrt{-\tilde{\lambda}}x - 3\tilde{\lambda})}{k(kx^2 + 3(\sqrt{-\tilde{\lambda}}x + \dot{x}))^2}. \tag{80}$$

The above integrating factors also satisfy the first equation of Eq. (33).

4 Noether's Theorem and Variational Symmetries

If the given second-order equation has a variational structure then one can also determine the symmetries which leave the action integral invariant. Such symmetries are called variational symmetries. Variational symmetries are important since they provide conservation laws via Noether's theorem [21]. In the following, we recall the method of finding variational symmetries [2, 23].

Noether's theorem states that whenever the action integral $S = \int L(t, x, \dot{x})dt$, where L is the Lagrangian, is invariant under the one parameter continuous group of transformations (7) then the solution of Euler's equation admit the conserved quantity [11, 14],

$$I = (\xi\dot{x} - \eta)\frac{\partial L}{\partial \dot{x}} - \xi L + f, \tag{81}$$

where f is an arbitrary function of t and x. The functions ξ, η and f can be determined from the equation

$$G\{L\} = -\dot{\xi} L + f, \tag{82}$$

where overdot denotes differentiation with respect to time and

$$G\{L\} = \xi \frac{\partial L}{\partial t} + \eta \frac{\partial L}{\partial x} + (\dot{\eta} - \dot{x}\dot{\xi}) \frac{\partial L}{\partial \dot{x}}. \tag{83}$$

Equation (83) can be derived by differentiating the Eq. (81) and simplifying the expressions in the resultant equation. Solving Eq. (83) one can obtain explicit expressions for Noether's symmetries ξ, η and the arbitrary function f. Now substituting these expressions into (81) one can get explicitly the associated integrals of motion.

To derive Noether's symmetries associated with the Lagrangian (2) let us substitute the expression (2) into (81). Doing so we get

$$\begin{aligned} &\eta \left(\frac{-9\tilde{\lambda}^3 x}{(k\dot{x} + \frac{k^2}{3}x^2 + 3\tilde{\lambda})^2} \right) + (\eta_t + \dot{x}\eta_x - \dot{x}(\xi_t + \dot{x}\xi_x)) \left(\frac{-27\tilde{\lambda}^3}{2k(k\dot{x} + \frac{k^2}{3}x^2 + 3\tilde{\lambda})^2} + \frac{3\tilde{\lambda}}{2k} \right) \\ &= -(\xi_t + \dot{x}\xi_x) \left(\frac{27\tilde{\lambda}^3}{2k^2} \left(\frac{1}{k\dot{x} + \frac{k^2}{3}x^2 + 3\tilde{\lambda}} \right) + \frac{3\tilde{\lambda}}{2k}\dot{x} - \frac{9\tilde{\lambda}^2}{2k^2} \right) + f_t + \dot{x} f_x. \end{aligned} \tag{84}$$

Now equating the coefficient of various powers of $\dot{x}$ to zero and solving the resultant equations we obtain three different forms of infinitesimal symmetries for ξ and η depending upon the sign and magnitude of $\tilde{\lambda}$. In the following, we discuss each one of the cases separately.

4.1 Case 1: $\tilde{\lambda} > 0$

Solving the determining equations with $\tilde{\lambda} > 0$, the associated vector fields turn out to be

$$\begin{aligned} X_1 &= \frac{1}{\tilde{\lambda}^{\frac{5}{2}}} \Bigg[\left(\sqrt{\tilde{\lambda}} \sin \sqrt{\tilde{\lambda}} t + \frac{k}{3} \cos \sqrt{\tilde{\lambda}} t x \right) \frac{\partial}{\partial t} \\ &\quad - \left(\frac{3\tilde{\lambda}}{k} + \frac{kx^2}{3} \right) \left(\sqrt{\tilde{\lambda}} \sin \sqrt{\tilde{\lambda}} t + \frac{k}{3} \cos \sqrt{\tilde{\lambda}} t x \right) \frac{\partial}{\partial x} \Bigg], \\ X_2 &= \frac{1}{\tilde{\lambda}^{\frac{5}{2}}} \Bigg[\left(\frac{k}{3} \sin \sqrt{\tilde{\lambda}} t x - \sqrt{\tilde{\lambda}} \cos \sqrt{\tilde{\lambda}} t \right) \frac{\partial}{\partial t} \\ &\quad - \left(\frac{3\tilde{\lambda}}{k} + \frac{kx^2}{3} \right) \left(\frac{k}{3} \sin \sqrt{\tilde{\lambda}} t x - \sqrt{\tilde{\lambda}} \cos \sqrt{\tilde{\lambda}} t \right) \frac{\partial}{\partial x} \Bigg], \\ X_3 &= \frac{1}{\tilde{\lambda}^{\frac{5}{2}}} \Bigg[\left(3\sqrt{\tilde{\lambda}} \sin 2\sqrt{\tilde{\lambda}} t + k \cos 2\sqrt{\tilde{\lambda}} t x \right) \frac{\partial}{\partial t} \end{aligned}$$

$$-\left(\left(\frac{k^2x^2}{3}-3\tilde{\lambda}\right)\cos 2\sqrt{\tilde{\lambda}}tx+2k\sqrt{\tilde{\lambda}}\sin 2\sqrt{\tilde{\lambda}}tx^2\right)\frac{\partial}{\partial x}\Bigg],$$

$$X_4=\frac{1}{\tilde{\lambda}^{\frac{5}{2}}}\Bigg[\cos\sqrt{\tilde{\lambda}}t\left(k\sin\sqrt{\tilde{\lambda}}tx-3\sqrt{\tilde{\lambda}}\cos\sqrt{\tilde{\lambda}}t\right)\frac{\partial}{\partial t}$$

$$-\left(\frac{k}{3}\sin\sqrt{\tilde{\lambda}}tx-\sqrt{\tilde{\lambda}}\cos\sqrt{\tilde{\lambda}}t\right)\left(3\sqrt{\tilde{\lambda}}\sin\sqrt{\tilde{\lambda}}tx+k\cos\sqrt{\tilde{\lambda}}tx^2\right)\frac{\partial}{\partial x}\Bigg],$$

$$X_5=\frac{\partial}{\partial t}. \tag{85}$$

Substituting each vector field into (81) we obtain the following integrals of motion

$$I_1=\frac{(3\dot{x}+kx^2)\cos\sqrt{\tilde{\lambda}}t+3\sqrt{\tilde{\lambda}}x\sin\sqrt{\tilde{\lambda}}t}{\alpha},\quad I_2=\frac{(3\dot{x}+kx^2)\sin\sqrt{\tilde{\lambda}}t-3\sqrt{\tilde{\lambda}}x\cos\sqrt{\tilde{\lambda}}t}{\alpha},$$

$$I_3=\frac{\left((3\dot{x}+kx^2)^2-9\tilde{\lambda}x^2\right)\sin 2\sqrt{\tilde{\lambda}}t-6x(3\dot{x}+kx^2)\sqrt{\tilde{\lambda}}\cos 2\sqrt{\tilde{\lambda}}t}{(\alpha)^2},$$

$$I_4=\frac{k^2((3\dot{x}+kx^2)^2-9k^2\tilde{\lambda}x^2)\cos 2\sqrt{\tilde{\lambda}}t+6k^2x(3\dot{x}+kx^2)\sqrt{\tilde{\lambda}}\sin 2\sqrt{\tilde{\lambda}}t-9\tilde{\lambda}(k^2x^2+6k\dot{x}+9\tilde{\lambda})}{(\alpha)^2},$$

$$I_5=\left(\frac{(3\dot{x}+kx^2)^2+9\tilde{\lambda}x^2}{(\alpha)^2}\right), \tag{86}$$

where $\alpha = 3k\dot{x} + k^2x^2 + 9\tilde{\lambda}$.

One can select two independent integrals of motions, I_1 and I_2 from the above. The remaining integrals of motions can be written in terms of the integrals of motion I_1 and I_2. For example, in the present case we get

$$I_3=2I_1I_2,\quad I_4=-1+2k^2I_1^2,\quad I_5=(I_1^2+I_2^2). \tag{87}$$

Using I_1 and I_2 we can construct the general solution in the form

$$x(t)=3\sqrt{\tilde{\lambda}}\left(\frac{I_1\sin\sqrt{\tilde{\lambda}}t-I_2\cos\sqrt{\tilde{\lambda}}t}{1-k(I_1\cos\sqrt{\tilde{\lambda}}t+I_2\sin\sqrt{\tilde{\lambda}}t)}\right). \tag{88}$$

The above solution is obviously equivalent to (73). Since Eq. (1) admits five Noether's symmetries for the case $\tilde{\lambda} > 0$ and so the Lagrangian (2) can be considered as a physically important Lagrangian from Quantum Mechanics point of view.

4.2 Case 2: $\tilde{\lambda} < 0$

Solving the determining equations with $\tilde{\lambda} < 0$, its associated symmetry vector fields turn out to be

$$X_1 = \frac{1}{\tilde{\lambda}^{\frac{5}{2}}}\left[e^{\sqrt{\tilde{\lambda}}t}\left(\left(\sqrt{\tilde{\lambda}}-\frac{kx}{3}\right)\frac{\partial}{\partial t}+\left(\frac{\sqrt{\tilde{\lambda}}}{k}-\frac{x}{3}\right)\left(3\tilde{\lambda}-\frac{k^2x^2}{3}\right)\frac{\partial}{\partial x}\right)\right],\ X_2 = \frac{\partial}{\partial t}$$
$$X_3 = \frac{1}{\tilde{\lambda}^{\frac{5}{2}}}\left[e^{-\sqrt{\tilde{\lambda}}t}\left(\left(\sqrt{\tilde{\lambda}}+\frac{kx}{3}\right)\frac{\partial}{\partial t}+\left(\frac{\sqrt{\tilde{\lambda}}}{k}+\frac{x}{3}\right)\left(3\tilde{\lambda}-\frac{k^2x^2}{3}\right)\frac{\partial}{\partial x}\right)\right],$$
$$X_4 = \frac{1}{\tilde{\lambda}^{\frac{5}{2}}}\left[e^{2\sqrt{\tilde{\lambda}}t}\left(\left(\sqrt{\tilde{\lambda}}-\frac{kx}{3}\right)\frac{\partial}{\partial t}+x\left(\tilde{\lambda}+\frac{k^2x^2}{9}-\frac{2}{3}kx\sqrt{\tilde{\lambda}}\right)\frac{\partial}{\partial x}\right)\right],$$
$$X_5 = \frac{1}{\tilde{\lambda}^{\frac{5}{2}}}\left[e^{-2\sqrt{\tilde{\lambda}}t}\left(\left(\sqrt{\tilde{\lambda}}+\frac{kx}{3}\right)\frac{\partial}{\partial t}-x\left(\tilde{\lambda}+\frac{k^2x^2}{9}+\frac{2}{3}kx\sqrt{\tilde{\lambda}}\right)\frac{\partial}{\partial x}\right)\right]. \tag{89}$$

Substituting each vector field into (81) we obtain the following integrals of motion,

$$I_1 = ke^{a_2t}\left(\frac{3\dot{x}-3a_2x+kx^2}{\alpha}\right),\ I_2 = ke^{-a_2t}\left(\frac{3\dot{x}+3a_2x+kx^2}{\alpha}\right),$$
$$I_3 = k^2e^{2a_2t}\left[\frac{(-6ka_2x^3+k^2x^4-18a_2x\dot{x}+9\dot{x}^2+x^2(-9\tilde{\lambda}+6k\dot{x}))}{(\alpha)^2}\right],$$
$$I_4 = k^2e^{-2a_2t}\left[\frac{(6ka_2x^3+k^2x^4+18a_2x\dot{x}+9\dot{x}^2+x^2(-9\tilde{\lambda}+6k\dot{x}))}{(\alpha)^2}\right],$$
$$I_5 = \frac{9\tilde{\lambda}^2}{2}\left(\frac{(3\dot{x}+kx^2)^2+9\tilde{\lambda}x^2}{(\alpha)^2}\right). \tag{90}$$

As in the previous case the integrals of motions, I_1 and I_2 are functionally independent from the rest. In other words the remaining integrals of motions can be written in terms of the integrals of motion I_1 and I_2:

$$I_3 = I_1^2,\quad I_4 = I_2^2,\quad I_5 = \frac{9\tilde{\lambda}^2}{2k^2}I_1I_2. \tag{91}$$

Using I_1 and I_2 we can construct the general solution in the form

$$x(t) = \frac{3a_2(I_1e^{2a_2t}-I_2)}{k(I_2+I_1e^{2a_2t}-2e^{a_2t})}. \tag{92}$$

The above solution is of front like nature and in this case also we have five Noether's symmetries. The underlying Lagrangian (2) is again a physically important one.

4.3 Case 3: $\tilde{\lambda} = 0$

One may note that in the limit $\tilde{\lambda} = 0$, Eq. (1) becomes the modified Emden equation/second order Ricatti equation which is another Liénard type system which possesses several interesting properties. Interestingly, this system also admits a time independent Lagrangian and Hamiltonian. In the following, we present the Noether's symmetries and their associated constants of motions.

The Lagrangian associated with the MEE equation is,

$$L = \frac{1}{k\dot{x} + \frac{k^2}{3}x^2}. \tag{93}$$

Solving the determining Eq. (83) with $\tilde{\lambda} = 0$, we get

$$\begin{aligned}
X_1 &= x\frac{\partial}{\partial t} - \frac{kx^2}{3}\frac{\partial}{\partial x}, \; X_2 = xt\frac{\partial}{\partial t} + \left(x^2 - \frac{ktx^3}{3}\right)\frac{\partial}{\partial x}, \; X_3 = \frac{\partial}{\partial t}, \\
X_4 &= \left(t - \frac{kt^2x}{2}\right)\frac{\partial}{\partial t} + \left(2x - ktx^2 + \frac{k^2t^2x^3}{6}\right)\frac{\partial}{\partial x}, \\
X_5 &= \left(\frac{k^2t^3x}{18} - \frac{kt^2}{6}\right)\frac{\partial}{\partial t} + \left(1 - \frac{2ktx}{3} + \frac{k^2t^2x^2}{6} - \frac{k^3t^3x^3}{54}\right)\frac{\partial}{\partial x}.
\end{aligned} \tag{94}$$

Substituting each vector field into (81) we obtain the following integrals of motion,

$$\begin{aligned}
I_1 &= t - \frac{3x}{kx^2 + 3\dot{x}}, \; I_2 = \frac{(-3x + ktx^2 + 3t\dot{x})^2}{(kx^2 + 3\dot{x})^2}, \\
I_3 &= \frac{-9k^2t^2x^3 + k^3t^3x^4 - 27x(2 + kt^2\dot{x}) + 6ktx^2(6 + kt^2\dot{x}) + 9t\dot{x}(6 + kt^2\dot{x})}{(kx^2 + 3\dot{x})^2}, \\
I_4 &= \frac{(18 - 6ktx + k^2t^2x^2 + 3kt^2\dot{x})}{(kx^2 + 3\dot{x})}, \; I_5 = \frac{6\dot{x} + kx^2}{(kx^2 + 3\dot{x})^2}.
\end{aligned} \tag{95}$$

One can easily check that out of the five integrals of motions two are independent and the remaining three can be expressed in terms of the first two, namely

$$I_2 = I_1^2, \quad I_3 = I_1 I_4, \quad I_5 = \frac{1}{9}(I_4 - kI_1^2). \tag{96}$$

We can construct a general solution of the form

$$x(t) = \frac{6(t - I_1)}{kt^2 - 2I_1kt + I_4}, \tag{97}$$

using I_1 and I_4. This case also admits five Noether's symmetries and so the Lagrangian (93) is physically important for $\tilde{\lambda} = 0$ in Eq. (1).

5 Telescopic Vector Fields

Telescopic vector fields are more general vector fields than the ones discussed so far. The Lie point symmetries, contact symmetries and λ-symmetries are all subcases of telescopic vector fields. A telescopic vector field can be considered as a λ-prolongation where the two first infinitesimals can depend on the first derivative of the dependent variable [10, 20, 26]. In the following, we briefly discuss the method of finding telescopic vector fields for a second-order ODE. We then present the telescopic vector fields for Eq. (1).

Let us consider the second-order Eq. (1). The vector field

$$\Omega^{(2)} = \xi\frac{\partial}{\partial t} + \eta\frac{\partial}{\partial x} + \zeta^{(1)}\frac{\partial}{\partial \dot{x}} + \zeta^{(2)}\frac{\partial}{\partial \ddot{x}} \tag{98}$$

is telescopic if and only if [26]

$$\xi = \xi(t, x, \dot{x}),\ \eta = \eta(t, x, \dot{x}),\ \zeta^{(1)} = \zeta^{(1)}(t, x, \dot{x}) \tag{99}$$

with $\zeta^{(2)}$ given by

$$\zeta^{(2)} = D[\zeta^{(1)}] - \phi D[\xi] + \frac{\zeta^{(1)} + \dot{x}D[\xi] - D[\eta]}{\eta - \dot{x}\xi}(\zeta^{(1)} - \phi\xi), \tag{100}$$

where ϕ is the given equation ($\ddot{x} = \phi(t, x, x\dot{x})$).

To prove that the telescopic vector fields are more general vector fields, let us introduce two functions g_1 and g_2 in the following forms, namely

$$g_1(t, x, \dot{x}) = \frac{\zeta^{(1)} + \dot{x}\xi_t - \eta_t + \dot{x}(\dot{x}\xi_x - \eta_x)}{\eta - \dot{x}\xi},\quad g_2(t, x, \dot{x}) = \frac{\dot{x}\xi_{\dot{x}} - \eta_{\dot{x}}}{\eta - \dot{x}\xi}. \tag{101}$$

We can rewrite the prolongations $\zeta^{(1)}$ and $\zeta^{(2)}$ using the above functions g_1 and g_2 as follows:

$$\zeta^{(1)} = D[\eta] - \dot{x}D[\xi] + (g_1 + g_2\phi)(\eta - \dot{x}\xi), \tag{102}$$

$$\zeta^{(2)} = D[\zeta^{(1)}] - \phi x D[\xi] + (g_1 + g_2\phi)(\zeta^{(1)} - \phi\xi). \tag{103}$$

The relationship between telescopic vector fields and previously considered vector fields can be given by the following expressions [10, 26],

$$\zeta^{(1)} = \eta^{(1)} + (g_1 + g_2\phi)(\eta - \dot{x}\xi), \tag{104}$$

$$\zeta^{(2)} = \eta^{(2)} + (g_1 + g_2\phi)(\zeta^{(1)} - \phi\xi). \tag{105}$$

In the above vector fields if we choose $g_1 = g_2 = 0$ and $\xi_{\dot{x}}^2 + \eta_{\dot{x}}^2 = 0$ we get the Lie point symmetries. The choice $g_1 = g_2 = 0$ and $\xi_{\dot{x}}^2 + \eta_{\dot{x}}^2 \neq 0$ gives the contact symmetries. To get λ-symmetries, we should choose $g_1 \neq 0$ and $\xi_{\dot{x}}^2 + \eta_{\dot{x}}^2 = 0$. As a consequence it can be considered as the more general vector field.

Hence the unknowns to be solved in Eq. (32) can also be $(\xi, \eta, \eta^{[\lambda,1]})$ by expressing λ in terms of $(\xi, \eta, \eta^{[\lambda,1]})$. In other words, if the given equation admits the telescopic vector field, then it satisfies the following invariance condition

$$\xi\frac{\partial\phi}{\partial t} + \eta\frac{\partial\phi}{\partial x} + \zeta^{(1)}\frac{\partial\phi}{\partial\dot{x}} - \zeta^{(2)} = 0. \tag{106}$$

In the above expression, ξ, η and $\zeta^{(1)}$ are three unknown functions which we need in order to write the telescopic vector fields of Eq. (1). Since the above expression has three unknowns, it is very difficult to find them systematically. For this purpose, we assume $\xi = 0$ and the remaining two unknown functions can be obtained in the following way. In this case, Eq. (106) turns out to be

$$\eta\frac{\partial\phi}{\partial x} + \zeta^{(1)}\frac{\partial\phi}{\partial\dot{x}} - \zeta^{(2)} = 0. \tag{107}$$

By assuming suitable ansatz for η and $\zeta^{(1)}$ we can find the telescopic vector fields associated with Eq. (1).

5.1 *Case 1:* $\tilde{\lambda} = 0$

For simplicity, first let us consider the case $\tilde{\lambda} = 0$. Assuming the ansatz

$$\eta = \frac{a_{01} + b_{01}\dot{x} + c_{01}\dot{x}^2}{(d_{01} + e_{01}\dot{x})^m}, \quad \zeta^{(1)} = \frac{a_{11} + b_{11}\dot{x} + c_{11}\dot{x}^2}{(d_{11} + e_{11}\dot{x})^n}, \tag{108}$$

for η and $\zeta^{(1)}$ and substituting them into Eq. (107) and solving the resultant expression we find the following telescopic vector fields for the case $\tilde{\lambda} = 0$:

$$\begin{aligned}
\Omega_1 &= \frac{9x}{(kx^2+3\dot{x})^2}\frac{\partial}{\partial x} + \frac{9\dot{x}-3kx^2}{(kx^2+3\dot{x})^2}\frac{\partial}{\partial\dot{x}} - \frac{18kx\dot{x}}{(kx^2+3\dot{x})^2}\frac{\partial}{\partial\ddot{x}},\\
\Omega_2 &= -\frac{18x(ktx^2+3t\dot{x}-3x)}{(kx^2+3\dot{x})^3}\frac{\partial}{\partial x} + \frac{6(kx^2-3\dot{x})(ktx^2+3t\dot{x}-3x)}{(kx^2+3\dot{x})^3}\frac{\partial}{\partial\dot{x}}\\
&\quad + \frac{36kx\dot{x}(ktx^2+3t\dot{x}-3x)}{(kx^2+3\dot{x})^3}\frac{\partial}{\partial\ddot{x}},\\
\Omega_3 &= \frac{-81t\dot{x}(ktx-2) - 27x(ktx(ktx-6)+12)}{(kx^2+3\dot{x})^3}\frac{\partial}{\partial x}\\
&\quad + \frac{9(-9kt^2\dot{x}^2 + kx^2(ktx(ktx-8)+18) - 18\dot{x})}{(kx^2+3\dot{x})^3}\frac{\partial}{\partial\dot{x}}
\end{aligned}$$

$$+\frac{18k(k^2tx^3(3t\dot{x}+x)-3kx(-3t^2\dot{x}^2+4tx\dot{x}+x^2)-9\dot{x}(t\dot{x}-3x))}{(kx^2+3\dot{x})^3}\frac{\partial}{\partial \ddot{x}},$$

$$\Omega_4=-\frac{18(ktx-3)}{(kx^2+3\dot{x})^2}\frac{\partial}{\partial x}+\frac{6k(x(ktx-6)-3t\dot{x})}{(kx^2+3\dot{x})^2}\frac{\partial}{\partial \dot{x}}+\frac{18k(kx(2t\dot{x}+x)-3\dot{x})}{(kx^2+3\dot{x})^2}\frac{\partial}{\partial \ddot{x}},$$

$$\Omega_5=\frac{18\dot{x}}{(kx^2+3\dot{x})^3}\frac{\partial}{\partial x}-\frac{2kx(kx^2+9\dot{x})}{(kx^2+3\dot{x})^3}\frac{\partial}{\partial \dot{x}}+\frac{2k(k^2x^4+6kx^2\dot{x}-9\dot{x}^2)}{(kx^2+3\dot{x})^3}\frac{\partial}{\partial \ddot{x}}. \quad (109)$$

The above telescopic vector fields also satisfy the invariance condition (106) with the choice $\tilde{\lambda}=0$. To find the solution from the above admitted telescopic vector fields, one has to follow the standard order-reduction procedure. Let us consider the telescopic vector field Ω_1. The corresponding Lagrange system can be written as

$$\frac{dt}{0}=\frac{dx}{\frac{9x}{(kx^2+3\dot{x})^2}}=\frac{d\dot{x}}{\frac{9\dot{x}-3kx^2}{(kx^2+3\dot{x})^2}}. \quad (110)$$

Solving the above set of equations, we get the characteristics as

$$u=t \text{ and } v=\frac{kx^2+3\dot{x}}{3x}. \quad (111)$$

From the above expression, we get $\frac{dv}{du}$ as

$$\frac{dv}{du}=-v^2. \quad (112)$$

Solution of the above equation is given by

$$v=\frac{1}{u-I_1}. \quad (113)$$

Substituting (111) into (113) and rewriting it, we get a first-order ODE

$$\dot{x}+\frac{x(I_1kx-ktx+3)}{3(I_1-t)}=0. \quad (114)$$

Integrating the above equation, we get the general solution of (1) for the choice $\tilde{\lambda}=0$ as

$$x(t)=\frac{6(I_1-t)}{-6I_2+2I_1kt-kt^2}, \quad (115)$$

where I_1 and I_2 are the integration constants.

5.2 Case 2: $\tilde{\lambda} > 0$

As we did in the previous case, here we obtain the following vector fields for the case $\tilde{\lambda} > 0$:

$$\begin{aligned}
\Omega_1 &= \frac{9(\sqrt{\tilde{\lambda}}kx\sin(\sqrt{\tilde{\lambda}}t) - 3\tilde{\lambda}\cos(\sqrt{\tilde{\lambda}}t))}{(\alpha)^2}\frac{\partial}{\partial x} \\
&+ \frac{3\sqrt{\tilde{\lambda}}\sin(\sqrt{\tilde{\lambda}}t)(9\tilde{\lambda} - k^2x^2 + 3k\dot{x}) + 18\tilde{\lambda}kx\cos(\sqrt{\tilde{\lambda}}t)}{(\alpha)^2}\frac{\partial}{\partial \dot{x}} \\
&+ \frac{9(\tilde{\lambda}\cos(\sqrt{\tilde{\lambda}}t)(3\tilde{\lambda} - k^2x^2 + 3k\dot{x}) - 2\sqrt{\tilde{\lambda}}kx(2\tilde{\lambda} + k\dot{x})\sin(\sqrt{\tilde{\lambda}}t))}{(\alpha)^2}\frac{\partial}{\partial \ddot{x}}, \\
\Omega_2 &= -\frac{9(\sqrt{\tilde{\lambda}}kx\cos(\sqrt{\tilde{\lambda}}t) + 3\tilde{\lambda}\sin(\sqrt{\tilde{\lambda}}t))}{(\alpha)^2}\frac{\partial}{\partial x} \\
&- \frac{3\sqrt{\tilde{\lambda}}\cos(\sqrt{\tilde{\lambda}}t)(9\tilde{\lambda} - k^2x^2 + 3k\dot{x}) - 18\tilde{\lambda}kx\sin(\sqrt{\tilde{\lambda}}t)}{(\alpha)^2}\frac{\partial}{\partial \dot{x}} \\
&+ \frac{9\tilde{\lambda}\sin(\sqrt{\tilde{\lambda}}t)(3\tilde{\lambda} - k^2x^2 + 3k\dot{x}) + 18\sqrt{\tilde{\lambda}}kx(2\tilde{\lambda} + k\dot{x})\cos(\sqrt{\tilde{\lambda}}t)}{(\alpha)^2}\frac{\partial}{\partial \ddot{x}}, \\
\Omega_3 &= \frac{18\sqrt{\tilde{\lambda}}(\alpha x\cos(2\sqrt{\tilde{\lambda}}t) - 3\sqrt{\tilde{\lambda}}(2kx^2 + 3\dot{x})\sin(2\sqrt{\tilde{\lambda}}t))}{(\alpha)^3}\frac{\partial}{\partial x} \\
&- \frac{54\tilde{\lambda}x\sin(2\sqrt{\tilde{\lambda}}t)d_1 - 6\sqrt{\tilde{\lambda}}\cos(2\sqrt{\tilde{\lambda}}t)a_1}{(\alpha)^3}\frac{\partial}{\partial \dot{x}} \\
&+ \frac{18(\sqrt{\tilde{\lambda}}x\cos(2\sqrt{\tilde{\lambda}}t)b_1 + \tilde{\lambda}\sin(2\sqrt{\tilde{\lambda}}t)c_1)}{(\alpha)^3}\frac{\partial}{\partial \ddot{x}}, \\
\Omega_4 &= \frac{36\sqrt{\tilde{\lambda}}k^2(kx\sin(\sqrt{\tilde{\lambda}}t) - 3\sqrt{\tilde{\lambda}}\cos(\sqrt{\tilde{\lambda}}t))((kx^2 + 3\dot{x})\cos(\sqrt{\tilde{\lambda}}t) + 3\sqrt{\tilde{\lambda}}x\sin(\sqrt{\tilde{\lambda}}t))}{(\alpha)^3}\frac{\partial}{\partial x} \\
&+ \frac{54\tilde{\lambda}k^2x\cos(2\sqrt{\tilde{\lambda}}t)d_1 + 18\tilde{\lambda}k^2x(9\tilde{\lambda} + k^2x^2 + 9k\dot{x}) + 6\sqrt{\tilde{\lambda}}k^2\sin(2\sqrt{\tilde{\lambda}}t)a_1}{(\alpha)^3}\frac{\partial}{\partial \dot{x}} \\
&- \frac{18\sqrt{\tilde{\lambda}}k^2(x\sin(2\sqrt{\tilde{\lambda}}t)b_1 - \sqrt{\tilde{\lambda}}\cos(2\sqrt{\tilde{\lambda}}t)c_1 + \sqrt{\tilde{\lambda}}e_1)}{(\alpha)^3}\frac{\partial}{\partial \ddot{x}}, \\
\Omega_5 &= -\frac{162\tilde{\lambda}\dot{x}}{(\alpha)^3}\frac{\partial}{\partial x} + \frac{18\tilde{\lambda}x(9\tilde{\lambda} + k^2x^2 + 9k\dot{x})}{(\alpha)^3}\frac{\partial}{\partial \dot{x}} - \frac{18\tilde{\lambda}e_1}{(\alpha)^3}\frac{\partial}{\partial \ddot{x}},
\end{aligned} \tag{116}$$

where $a_1 = (-k^3x^4 + 9k(3\tilde{\lambda}x^2 + \dot{x}^2) + 27\tilde{\lambda}\dot{x})$, $b_1 = (-9\tilde{\lambda}^2 + 2k^3x^2\dot{x} + k^2(7\tilde{\lambda}x^2 + 6\dot{x}^2) + 3\tilde{\lambda}k\dot{x})$, $c_1 = (-k^3x^4 + 6k^2x^2\dot{x} + 15\tilde{\lambda}kx^2 + 9k\dot{x}^2 + 9\tilde{\lambda}\dot{x})$, $d_1 = k(kx^2 + \dot{x}) - 3\tilde{\lambda}$ and $e_1 = (k^3x^4 + 6k^2x^2\dot{x} + 9\tilde{\lambda}kx^2 - 9k\dot{x}^2 - 9\tilde{\lambda}\dot{x})$. The above telescopic vector fields also satisfy the invariance condition (106). To find the solution from the above admitted telescopic vector fields, one has to follow the standard order-reduction procedure. Let us consider the telescopic vector field Ω_5. The corresponding Lagrange system can be written as

$$\frac{dt}{0} = \frac{dx}{-\frac{162\lambda\dot{x}}{(k^2x^2+3k\dot{x}+9\lambda)^3}} = \frac{d\dot{x}}{\frac{18\lambda x(k^2x^2+9k\dot{x}+9\lambda)}{(k^2x^2+3k\dot{x}+9\lambda)^3}}. \tag{117}$$

Solving the above set of equations, we get the characteristics as

$$u = t \text{ and } v = \log\left(\frac{81\sqrt{3}(k^2x^2 + 3k\dot{x} + 9\lambda)^9}{\sqrt{(k^2x^2 + 6k\dot{x} + 9\lambda)^9}}\right). \tag{118}$$

From the above expression, we get $\frac{dv}{dt} = 0$. So the function v itself acts as a first integral. Then the integral I_1 takes the form

$$I_1 = \frac{81\sqrt{3}(k^2x^2 + 3k\dot{x} + 9\lambda)^9}{\sqrt{(k^2x^2 + 6k\dot{x} + 9\lambda)^9}}. \tag{119}$$

Rewriting the above expression for $\dot{x}$ and integrating it we obtain the general solution as in Eq. (88).

5.3 Case 3: $\tilde{\lambda} < 0$

For the case $\tilde{\lambda} < 0$, we get the telescopic vector fields by following the procedure discussed in the case $\tilde{\lambda} = 0$. The telescopic vector fields are given by

$$\begin{aligned}
\Omega_1 &= -\frac{9ke^{a_2t}(3\tilde{\lambda} + a_2kx)}{(\alpha)^2}\frac{\partial}{\partial x} \\
&\quad + \frac{3ke^{a_2t}(9(-\tilde{\lambda})^{3/2} + a_2k^2x^2 + 6\tilde{\lambda}kx - 3a_2k\dot{x})}{(\alpha)^2}\frac{\partial}{\partial \dot{x}} \\
&\quad + \frac{9\tilde{\lambda}ke^{a_2t}(9\tilde{\lambda}^2 - k^3x^2(a_2x + 2\dot{x}) + k^2x(9a_2\dot{x} - 7\tilde{\lambda}x) + 3\tilde{\lambda}k(5a_2x + 3\dot{x}))}{(3\tilde{\lambda} + a_2kx)(\alpha)^2}\frac{\partial}{\partial \ddot{x}}, \\
\Omega_2 &= \frac{9ke^{a_2(-t)}(a_2kx - 3\tilde{\lambda})}{(\alpha)^2}\frac{\partial}{\partial x} \\
&\quad + \frac{3ke^{a_2(-t)}(9a_2\tilde{\lambda} + a_2(-k^2)x^2 + 6\tilde{\lambda}kx + 3a_2k\dot{x})}{(\alpha)^2}\frac{\partial}{\partial \dot{x}} \\
&\quad + \frac{9\tilde{\lambda}ke^{a_2(-t)}(9\tilde{\lambda}^2 + k^3x^2(a_2x - 2\dot{x}) - k^2x(7\tilde{\lambda}x + 9a_2\dot{x}) + 3\tilde{\lambda}k(3\dot{x} - 5a_2x))}{(3\tilde{\lambda} - a_2kx)(\alpha)^2}\frac{\partial}{\partial \ddot{x}}, \\
\Omega_3 &= -\frac{18k^2e^{2a_2t}(a_2k^2x^3 + 6\tilde{\lambda}kx^2 + 3a_2kx\dot{x} + 9(-\tilde{\lambda})^{3/2}x + 9\tilde{\lambda}\dot{x})}{(\alpha)^3}\frac{\partial}{\partial x} \\
&\quad + \frac{6k^2e^{2a_2t}(a_2k^3x^4 + 9\tilde{\lambda}k^2x^3 - 9k(3a_2\tilde{\lambda}x^2 - \tilde{\lambda}x\dot{x} + a_2\dot{x}^2) - 27\tilde{\lambda}(\tilde{\lambda}x + a_2\dot{x}))}{(\alpha)^3}\frac{\partial}{\partial \dot{x}} \\
&\quad + \frac{18k^2e^{2a_2t}(k^3x^3(2a_2\dot{x} - \tilde{\lambda}x) + k^2xb_2 + 3\tilde{\lambda}kc_2 + 9\tilde{\lambda}^2(\dot{x} - a_2x))}{(\alpha)^3}\frac{\partial}{\partial \ddot{x}}, \\
\Omega_4 &= \frac{18k^2e^{-2a_2t}(a_2k^2x^3 + 3kx(a_2\dot{x} - 2\tilde{\lambda}x) - 9\tilde{\lambda}(a_2x + \dot{x}))}{(\alpha)^3}\frac{\partial}{\partial x} \\
&\quad + \frac{6k^2e^{-2a_2t}(-a_2k^3x^4 + 9\tilde{\lambda}k^2x^3 + 9k(3a_2\tilde{\lambda}x^2 + \tilde{\lambda}x\dot{x} + a_2\dot{x}^2) + 27\tilde{\lambda}(a_2\dot{x} - \tilde{\lambda}x))}{(\alpha)^3}\frac{\partial}{\partial \dot{x}}
\end{aligned}$$

$$+\frac{18k^2e^{-2a_2t}(-k^3x^3(\tilde{\lambda}x+2a_2\dot{x})+k^2xd_2+3\tilde{\lambda}ke_2+9\tilde{\lambda}^2(a_2x+\dot{x}))}{(\alpha)^3}\frac{\partial}{\partial\ddot{x}},$$

$$\Omega_5=-\frac{729\tilde{\lambda}^3\dot{x}}{(\alpha)^3}\frac{\partial}{\partial x}+\frac{81\tilde{\lambda}^3x(9\tilde{\lambda}+k^2x^2+9k\dot{x})}{(\alpha)^3}\frac{\partial}{\partial\dot{x}}-\frac{81\tilde{\lambda}^3e_1}{(\alpha)^3}\frac{\partial}{\partial\ddot{x}}, \tag{120}$$

where $a_2=\sqrt{-\tilde{\lambda}}$, $b_2=(7a_2\tilde{\lambda}x^2+6\tilde{\lambda}x\dot{x}+6a_2\dot{x}^2)$, $c_2=(5\tilde{\lambda}x^2+a_2x\dot{x}+3\dot{x}^2)$, $sd_2=(7(-\tilde{\lambda})^{3/2}x^2+6\tilde{\lambda}x\dot{x}-6a_2\dot{x}^2)$, $e_2=(5\tilde{\lambda}x^2-a_2x\dot{x}+3\dot{x}^2)$. Here also one can check that the above telescopic vector fields satisfy the invariance condition (106). To find the solution from the above admitted telescopic vector fields, one has to follow the standard order-reduction procedure. Let us consider the telescopic vector field Ω_1. The corresponding Lagrange system can be written as

$$\frac{dt}{0}=\frac{dx}{-\frac{9ke^{\sqrt{-\lambda}t}(k\sqrt{-\lambda}x+3\lambda)}{(k^2x^2+3k\dot{x}+9\lambda)^2}}=\frac{d\dot{x}}{\frac{3ke^{\sqrt{-\lambda}t}(k^2\sqrt{-\lambda}x^2+6k\lambda x-3k\sqrt{-\lambda}\dot{x}+9(-\lambda)^{3/2})}{(k^2x^2+3k\dot{x}+9\lambda)^2}}. \tag{121}$$

Solving the above set of equations, we get the characteristics as

$$u=t \text{ and } v=\frac{k\sqrt{-\lambda}x^2+3\lambda x+3\sqrt{-\lambda}\dot{x}}{9\sqrt{-\lambda}\lambda-3k\lambda x}. \tag{122}$$

From the above expression, we get $\frac{dv}{du}$ as

$$\frac{dv}{du}=-3\sqrt{-\lambda}\left(\frac{kv^2}{3}+\frac{v}{3}\right). \tag{123}$$

Solution of the above equation is given by

$$v=\frac{e^{I_1}}{e^{\sqrt{-\lambda}u}-e^{I_1}k}. \tag{124}$$

Substituting (122) into (124) and rewriting it we get a first-order ODE,

$$\dot{x}+\frac{e^{I_1}\sqrt{-\lambda}(k^2x^2+9\lambda)-xe^{\sqrt{-\lambda}t}(k\sqrt{-\lambda}x+3\lambda)}{3\sqrt{-\lambda}(e^{\sqrt{-\lambda}t}-e^{I_1}k)}=0. \tag{125}$$

Integrating the above equation, we get the general solution of (1) for the choice $\tilde{\lambda}<0$ as

$$x(t)=\frac{3\sqrt{-\lambda}(-e^{I_1}k+2c_1e^{2\sqrt{-\lambda}t})}{k(2c_1e^{2\sqrt{-\lambda}t}+e^{I_1}k-2e^{\sqrt{-\lambda}t})}, \tag{126}$$

where I_1 and I_2 are the integration constants. Obviously the solution (126) can be rewritten in the form (92).

6 Conclusion

In this paper, we have reviewed four different kinds of symmetries for the Liénard type nonlinear oscillator Eq. (1). It has already been shown that this equation exhibits three different kinds of dynamics depending upon the sign of the parameter $\tilde{\lambda}$. Based on this earlier result we have divided our analysis into three categories while studying the symmetries of this equation. To begin with, we have considered Lie point symmetries of this equation. We have derived the general solution for all the three regimes by considering a vector field in each one of the cases. We then considered λ-symmetries approach to this equation. As we noted earlier, we carried out this calculations for the $\tilde{\lambda} = 0$ case and demonstrated the applicability of λ-symmetries approach in establishing the integrability of this equation. We have then studied the Noether's symmetries of (1) for the parametric choices $\tilde{\lambda} > 0$, $\tilde{\lambda} < 0$ and $\tilde{\lambda} = 0$. The underlying Lagrangian is of non-standard type. However in all the three cases, we found maximal number (five) of Noether's symmetries for the Lagrangian (2). Recently it has been proposed that the physical Lagrangian for a second order differential equation should be the one which admits highest possible number of Noether's symmetries. Our results indicate that even though the Lagrangian is of nonstandard type it can be considered as a physical Lagrangian since it admits maximal number of symmetries. Finally, we have constructed telescopic vector fields for Eq. (1) in the parametric regimes $\tilde{\lambda} > 0$, $\tilde{\lambda} < 0$ and $\tilde{\lambda} = 0$. The method of finding general solution from telescopic vector fields is also explained. Thus we have shown the utility of symmetry analysis in solving the nonlinear ODEs of Liénard type.

Acknowledgements RMS acknowledges the University Grants Commission (UGC), New Delhi, India, for the award of a Dr. D. S. Kothari Post Doctoral Fellowship [Award No. F.4-2/2006 (BSR)/PH/17-18/0004]. The work of M.S. forms part of CSIR research project under Grant No. 03(1397)/17/EMR-II. The work of V.K.C. was supported by the SERB-DST Fast Track scheme for young scientists under Grant No. YSS/2014/000175. The work of M.L. is supported by a DST-SERB Distinguished Fellowship program.

References

1. Bhuvaneswari, A., Kraenkel, R.A., Senthilvelan, M.: Nonlinear Anal. Real World Appl. **13**, 1102–1114 (2012)
2. Bluman, G.W., Anco, S.C.: Symmetries and Integration Methods for Differential Equations. Springer, New York (2002)
3. Chandrasekhar, S.: An Introduction to the Study of Stellar Structure. Dover, New York (1957)
4. Chandrasekar, V.K., Senthilvelan, M., Lakshmanan, M.: Phys. Rev. E **72**, 066203 (2005)
5. Chandrasekar, V.K., Senthilvelan, M., Lakshmanan, M.: Proc. R. Soc. Lond. A **461**, 2451–2476 (2005)
6. Chandrasekar, V.K., Pandey, S.N., Senthilvelan, M., Lakshmanan, M.: Chaos, Solitons Fractals **26**, 1399–1406 (2005)
7. Chandrasekar, V.K., Pandey, S.N., Senthilvelan, M., Lakshmanan, M.: J. Phys. A Math. Gen. **39**, L69–L76 (2006)

8. Chithiika Ruby, V., Senthilvelan, M., Lakshmanan, M.: J. Phys. A Math. Theor. **45**, 382002 (2012)
9. Choudhuri, A.: On the symmetries of the modified Emden-type equation. arXiv:1304.5826v1
10. Gaeta, G.: J. Nonlinear Math. Phys. **16**, 107 (2009)
11. Gelfand, I.M., Fomin, S.V.: Calculus of Variations Translator. In: Silverman, R.A. (ed.). Prentice Hall, Englewood Cliffs (1963)
12. Gubbiotti, G., Nucci, M.C.: J. Nonlinear Math. Phys. **21**, 248–264 (2014)
13. Hydon, P.E.: Symmetry Methods for Differential Equations: A Beginnner's Guide. Cambridge University Press, Cambridge (2000)
14. Lutzky, M.: J. Phys. A Math. Gen. **11**, 249–258 (1978)
15. Mohanasubha, R., Sabiya Shakila, M.I., Senthilvelan, M.: Commun. Nonlinear. Sci. **19**, 799–806 (2014)
16. Muriel, C., Romero, J.L.: λ-symmetries and linearization of ordinary differential equations through nonlocal transformations. Preprint
17. Muriel, C., Romero, J.L.: IMA J. Appl. Math. **66**, 111 (2001)
18. Muriel, C., Romero, J.L.: Theor. Math. Phys. **133**, 1565 (2002)
19. Muriel, C., Romero, J.L.: J. Phys. A Math. Theor. **42**, 365207 (2009)
20. Muriel, C., Romero, J.L.: SIGMA **8**, 106 (2012)
21. Noether, E.: Gttingen Nachrichten Mathematik-physik Klasse **2**, 235 (1918)
22. Nucci, M.C., Tamizhmani, K.M.: J. Nonlinear Math. Phys. **17**, 167 (2010)
23. Olver, P.J.: Applications of Lie Groups to Differential Equations. Springer, New York (1986)
24. Pandey, S.N., Bindu, P.S., Senthilvelan, M., Lakshmanan, M.: J. Math. Phys. **50**, 102701 (2009)
25. Polat, G.G., Özer, T.: Nonlinear Dyn. **85**, 1571–1595 (2016)
26. Pucci, E., Saccomandi, G.: J. Phys. A Math. Gen. **35**, 6145 (2012)
27. Yasar, E.: Mathematical Problems in Engineering, Article ID 916437, 10 pp. (2011)

Symmetries of Equations with Nonlocal Terms

Sergey V. Meleshko

Abstract An approach for applying group analysis to equations with nonlocal terms is given in the presentation. Similar to the theory of partial differential equations, for invariant solutions of equations with nonlocal terms the number of the independent variables is reduced. The presentation consists of reviewing results obtained by the author with his colleagues related with applications of the group analysis to equations with nonlocal terms such as: integro-differential equations, delay differential equations and stochastic differential equations. The proposed approach can also be applied for defining a Lie group of equivalence, contact and Lie–Bäcklund transformations for equations with nonlocal terms. The presentation is devoted to review the results where the author took a part.

Keywords Lie group · Symmetry · Invariant solution · Integro-differential equation · Delay differential equation · Stochastic differential equations

1 Introduction

Equations describing real phenomena in mathematical modelling take various forms, such as ordinary differential equations, partial differential equations, integro-differential equations, functional differential equations and many others. The algorithmic approach of group analysis was developed especially for differential equations. Applying it to equations having nonlocal terms presents some difficulties. The main ones of these arise from the nonlocal terms.

In applications of group analysis to equations with nonlocal operators it is necessary to use successive steps, as for partial differential equations. The first step involves constructing an admitted Lie group. Since the definition of an admitted Lie group given for partial differential equations cannot be applied to equations with

S. V. Meleshko (✉)
School of Mathematics, Institute of Science, Suranaree University of Technology, Nakhon Ratchasima, Thailand
e-mail: sergey@math.sut.ac.th

V. G. Kac et al. (eds.), *Symmetries, Differential Equations and Applications*,
Springer Proceedings in Mathematics & Statistics 266,
https://doi.org/10.1007/978-3-030-01376-9_6

nonlocal terms, this concept requires further investigation. Notice that even for partial differential equations the notion of an admitted Lie group needs to be discussed: there are three definitions of the admitted Lie group. The first part of the presentation is devoted to a discussion of these definitions. This discussion assists in establishing a definition of an admitted Lie group for differential equations with nonlocal terms.

As for partial differential equations, an admitted Lie group of equations with nonlocal terms is a Lie group satisfying determining equations. In contrast to partial differential equations the admitted Lie group does not have the property of mapping any solution into a solution of the same equations, although the method developed for constructing the determining equations used this property. In practice the algorithm for obtaining determining equations is no more difficult than for partial differential equations. The main difficulty consists of solving the determining equations. Since this depends on the properties of the Cauchy problem, the method of solving determining equations also depends on the nonlocal equations studied.

Because the method and some its details are given in [1, 2], the present manuscript is devoted to review the results obtained by the author with his colleagues after 2010. It should be noted that the review does not include results of other authors. References of papers of other authors can be found in the original publications.

1.1 Short Review of the Approaches

As mentioned the main difficulty in applications of group analysis to integro-differential equations arises from the integral (nonlocal) terms present in these equations. There are several heuristic ways for overcoming this difficulty. Among these ways the following are pointed out:

(1) finding a representation of an admitted group or a solution on the basis of a priory assumptions;
(2) studying a system of moments – the method of moments;
(3) transforming the original equations into differential ones.

The first approach supposes an a priori choice of the form of symmetries or solutions on the basis of some assumptions. This approach is the simplest and the most efficient. For example, the well-known BKW-solution of the Boltzmann equation was found in this way. For the Boltzmann equation this approach was also applied in [3, 4]. The main problem in this approach is to discover a representation of an admitted group (or solution).

In the second approach (the method of moments) the original system of integro-differential equations is reduced to an infinite system of differential equations (system of moment equations). For a finite number N of equations of this system, containing a finite number of terms, the classical group analysis (for differential equations) is applied. Then the process of taking a limit $N \to \infty$ is carried out. The first application of this approach for finding an admitted Lie group was done in [5], and then it was used for one of the models of the Boltzmann equation in [6, 7]. There are some problems in the application of this approach. One of them is that, for some equations, the construction of the moment system is impossible. Another problem with the moment system is the infinite number of equations that are involved.

In the third approach, as previously, initial integro-differential equations are transformed into differential equations. After that a classical algorithm of group analysis is applied to the differential equations. In this way there are the same problems as in the previous approaches.

For a complete description of group properties of equations with nonlocal terms it is necessary to use successive approaches of group analysis: constructing determining equations and finding their solutions. Such an approach for integro-differential equations was started developing in [3, 8]. Details of this developing are summarised in [2]. The present paper is devoted to discussion of our results obtained after publishing [2].

1.2 Revisiting Group Analysis of PDEs

For understanding the method for finding admitted Lie group of equations with nonlocal terms and compare it with the classical method of finding admitted Lie group of partial differential equations it would be useful to revisit the classical algorithm.

Consider differential equations

$$\Phi(x, u, p) = 0. \tag{1}$$

Here and further $x = (x_1, x_2, \ldots, x_n)$ is the set of independent variables, $u = (u^1, u^2, \ldots, u^m)$ is the set of dependent variables, $\alpha = (\alpha_1, \alpha_2, \ldots, \alpha_n)$ is a multiindex, p is the vector of the partial derivatives $p_\alpha = \frac{\partial^{|\alpha|}u}{\partial x_1^{\alpha_1}\partial x_2^{\alpha_2}\ldots\partial x_n^{\alpha_n}}$. For the multiindex α the following notations are used $|\alpha| = \alpha_1 + \alpha_2 + \cdots + \alpha_n, \alpha, j = (\alpha_1, \ldots, \alpha_{j-1}, \alpha_j + 1, \alpha_{j+1}, \ldots, \alpha_n)$, for $\alpha = 0$ one has $p_0^j = u^j$; for $\alpha_j = 1$ and $\alpha_i = 0,\ (i \neq j)$ it is denoted by $\alpha = j$.

In one of definitions, a one-parameter Lie group of transformations

$$\bar{x} = f^x(x, u; a), \quad \bar{u} = f^u(x, u; a) \tag{2}$$

with its infinitesimal generator

$$X = \zeta^{u^j}(x, u)\partial_{u^j} + \xi^{x_i}(x, u)\partial_{x_i}.$$

is called admitted if it satisfies the equations:

$$X_L \Phi_{|S} = 0, \tag{3}$$

where X_L is the prolongation of X:

$$X_L = X + \zeta^{p^j_\alpha}\partial_{p^j_\alpha}.$$

with the prolongation formulae

$$\zeta^{p^j_{\alpha,k}} = \zeta^{p^j_\alpha} - p^j_{\alpha,i} D_{x_k}\xi^{x_i}, \quad (i = 1, 2, \ldots, n;\ \ j = 1, 2, \ldots, m;\ \ |\alpha| = 0, 1, \ldots, N-1).$$

It is necessary to clarify here the following two questions.

1. How were the determining Eq. (3) found?
2. What is a meaning of $|S$.

1.2.1 Derivation of Determining Eq. (3)

Let $u = u_0(x)$ be a solution of Eq. (1). Applying the Lie group of transformations, one obtains the transformed function

$$u_a(\bar{x}) = f^u(g(\bar{x}, a), u_0(g(\bar{x}, a)), a),$$

where the function $x = g(\bar{x}, a)$ is a solution of the equations

$$\bar{x} = f^x(x, u_0(x), a).$$

Assuming that the transformed function is a solution of the same system of differential equations, one requires that

$$\Phi\,(\bar{x}, u_a(\bar{x}), p_a(\bar{x})) = 0, \tag{4}$$

where $p_a(\bar{x})$ are derivatives of the transformed function $u_a(\bar{x})$. An alternative form of Eq. (4) is

$$\Phi\left(f^x(x, u_0(x), a),\, f^u(x, u_0(x), a),\, f^p(x, u_0(x), p_0(x), a)\right) = 0, \tag{5}$$

where the functions f^p are obtained by prolonging the Lie group (2) on the derivatives p. Differentiating the latter form (5) with respect to the group parameter a, and setting $a = 0$, one derives the classical form of determining Eq. (3)

$$X_L\Phi\,(x, u_0(x), p_0(x)) = 0. \tag{6}$$

Whereas differentiating Eq. (4) with respect to the group parameter a, and setting $a = 0$, one derives the alternative form of determining Eq. (3)

$$X_{LB}\Phi\,(x, u_0(x), p_0(x)) = 0, \tag{7}$$

where the operator

$$X_{LB} = X_L - \xi^{x_i} D_{x_i}$$

is a canonical Lie–Bäcklund operator equivalent to the generator X_L.

1.2.2 Discussion of the Meaning $|S$

According to the derivation of the determining Eq. (3) or (7) the sign $|S$ means that the determining equations are satisfied for any solution of the equations $\Phi(x, u, p) = 0$.

For many systems of differential equations (for example, Cauchy–Kovalevskaya type) this condition coincides with the meaning that the determining equations are considered on the manifold defined by the equations $\Phi(x, u, p) = 0$ or their prolongations: for the determining Eq. (3) the manifold is $\Phi(x, u, p) = 0$; for the determining Eq. (7) the manifold is defined by $\Phi(x, u, p) = 0$ and $D_{x_i}\Phi(x, u, p) = 0$, $(i = 1, 2, \ldots, n)$.

If a system of differential equations (1) is not involutive it can produce new equations. If one requires to derive determining equations equivalent to the Eq. (3) or (7), then the new equations has to be added to the original system of equations.[1] The analysis of a Cauchy problem allows one to simplify determining equations by splitting them with respect to parameter derivatives.

2 Definition of Admitted Lie Group for Equations with Nonlocal Terms

Let

$$\Phi(x, u, p) = 0$$

be a system of equations with nonlocal terms. For deriving the determining equations for the latter equations it is more convenient to use the alternative to classical approach, where the variables $(\bar{x}, a)$ are considered as the independent variables,

$$\Phi\ (\bar{x}, u_a(\bar{x}), u_a(\bar{x})) = 0. \tag{8}$$

As for partial differential equations, differentiating the latter equations with respect to the group parameter a, and setting $a = 0$, one derives the equations:

$$X_{LB}\Phi\ (x, u_0(x), p_0(x)) = 0.$$

[1]One of the well-known examples of noninvolutive system of partial differential equations is the system of the Navier–Stokes equations. Although it should be mentioned that the solution of the determining equations of the original system of the Navier–Stokes equations and of the system extended by the equation making the system of the Navier–Stokes equations to be involutive do not change the admitted group.

Formally we write them as

$$X_{LB}\Phi_{|S} = 0, \tag{9}$$

where derivatives in the operator X_{LB} are considered in the Frèchet sense, and $|S$ means that the equations $X_{LB}\Phi = 0$ have to be satisfied for any solution of the equations $\Phi = 0$.

In the process of deriving the determining Eq. (9) it was assumed that all steps defined above are feasible. Whereas the computation of the determining Eq. (9) does not require this.

Definition A Lie group of transformations (2) satisfying the determining Eq. (9) is called a Lie group admitted by equations with nonlocal terms $\Phi(x, u, p) = 0$.

Example (Barba's equation) Here is an example demonstrating computation of the determining equations.

Consider the equation

$$y(x)y'(x) = y\left(y(x)\right). \tag{10}$$

The Lie group defined by the generator

$$X_L = \xi(x, y)\partial_x + \eta(x, y)\partial_y$$

is admitted by Eq. (10) if it satisfies the determining equation

$$\begin{array}{c} cX_{LB}\left(y(x)y'(x) - y\left(y(x)\right)\right) = \left(\eta(x, y(x)) - y'(x)\xi(x, y(x))\right)y'(x) \\ +y(x)(\eta_x(x, y(x)) + y'(x)\eta_y(x, y(x)) - y''(x)\xi(x, y(x)) \\ -y'(x)(\xi_x(x, y(x)) + y'(x)\xi_y(x, y(x)))) - (\eta(x, y(y(x))) - y'(y(x))\xi(x, y(y(x))) \\ +y'(y(x))\left(\eta(x, y(x)) - y'(x)\xi(x, y(x))\right)\right). \end{array}$$

3 Methods for Solving Determining Equations

Determining Eq. (9) are still nonlocal. In contrast to differential equations there are no general algorithms of their solving. As for partial differential equations, determining equations of some types of nonlocal equations can be simplified by splitting them. The splitting depends not only on a type of equations, but also on their form. In particular, splitting delay differential equations is similar to the splitting the determining equations of partial differential equations, whereas for some integro-differential equations the splitting cannot be applied. Similar to partial differential equations the algorithm for splitting is defined by the existence of a solution of a Cauchy problem. Moreover, for equations for which splitting cannot be applied, a choice of arbitrary elements in solving a Cauchy problem can assist in their solving.

In this section we demonstrate several methods of solving determining equations of different types of equations with nonlocal terms.

3.1 Expanding Coefficients of an Admitted Generator in Taylor Series

Usually the coefficients of admitted generators are assumed to be sufficiently smooth. In particular, one can suppose that they are analytic. Using a particular class of initial data and the latter assumption allow one to expand determining Eq. (9) in Taylor series. This provides equations for coefficients of the Taylor series. Solving these equations, one finds the general solution of the determining equations.

First application of the above presented method was given in [3],[2] where the Fourier image of the spatially homogeneous and isotropic Boltzmann equation was analyzed. Later this method was applied to different population balance equations [9, 10].

3.2 Using Arbitrariness of Integral Terms

Arbitrariness of the initial data of a Cauchy problem allows one to use their for splitting integro-differential equations. First application of this method was used in analysis of group properties of one-dimensional motion of viscoelastic media [11]. Recently this method was applied for group analysis of integro-differential equations describing stress relaxation behavior of one-dimensional viscoelastic materials [12], and then equations of a linear thermoviscoelasticity [13, 14].

3.3 A Method of Preliminary Group Classification

The classes of equations arising in science usually have undefined functions (arbitrary elements). The presence of these functions in the equations requires group classification with respect to them. The complete solution of the group classification problem is a nontrivial task. Nevertheless, for many equations with vanishing arbitrary elements the group analysis method has been applied. This analysis can assist in group classification of the equations with nonvanishing arbitrary elements. In [15] an approach for using this approach for the group classification of equations with nonvanishing arbitrary elements was proposed. The proposed approach was demonstrated not only by partial differential equations, but also by integro-differential equations and delay differential equations.

[2]Details can be found in [1, 2].

4 Symmetries of Integro-Differential Equations

Initially the discussed method was considered for integro-differential equations.

4.1 The Boltzmann Equation and Its Models

First application of solving determining Eq. (9) was presented in [3], where the Fourier image of the spatially homogeneous and isotropic Boltzmann equation [16] was considered:

$$\varphi_t(x,t) + \varphi(x,t)\varphi(0,t) = \int_0^1 \varphi(xs,t)\varphi(x(1-s),t)\,ds.$$

Using particular class of initial values, the general solution of the determining equation was found. Detailed review of the results related with the applications of the group analysis to the Boltzmann equation one can find in the recent publications [17, 18].

4.2 Population Balance Equations

Many chemical processes, including crystallization, aerosol formation, polymerization, and growth of cell populations, are best described by the population balance equation (PBE).

In [19], the following one-dimensional homogeneous population balance equation, used for batch crystallization units, was considered

$$\begin{aligned} c\frac{\partial f}{\partial t} + c^g\frac{\partial f}{\partial L} &= 0, \\ \frac{\partial c}{\partial t} &= -c^g\int_0^\infty fL^2\,dL. \end{aligned} \tag{11}$$

where t denotes the time, c is the solution concentration, L is an internal coordinate, the characteristic length of the particle (it can also represent age, composition, or other characteristics of an entity in a distribution depending on the system being modelled, although this may alter the mass balance equation), $f(L,t)$ is the probability distribution representing the number concentration of particles of a particular size, L, at the time t (this is commonly known as the population density).

It was found that the infinitesimal generators admitted by Eq. (11) form the three-dimensional Lie algebra L_3 spanned by the generators

$$X_1 = L\frac{\partial}{\partial L} + \left(\frac{1}{g} - 4\right) f\frac{\partial}{\partial f} + \frac{1}{g}c\frac{\partial}{\partial c}, \quad X_2 = t\frac{\partial}{\partial t} - \frac{1}{g}f\frac{\partial}{\partial f} - \frac{1}{g}c\frac{\partial}{\partial c}, \quad X_3 = \frac{\partial}{\partial t}. \tag{12}$$

All invariant solutions determined by this Lie algebra were considered in [19]. Analysis of the reduced equations is provided there.

In [9, 10], the PBE for continuous systems involving aggregation and crystal growth with one internal coordinate $\bar{x}$ and one external coordinate y is studied

$$\frac{\partial f(x,y,t)}{\partial t} = -\frac{\partial}{\partial x}[Gf(x,y,t)] - \frac{\partial}{\partial y}[Zf(x,y,t)]$$
$$+\frac{1}{2}\int_0^x K(x-z,z)f(x-z,y,t)f(z,y,t)\,dz - f(x,y,t)\int_0^\infty K(x,z)f(z,y,t)\,dz, \tag{13}$$

where t denotes the time, f is the one-dimensional population density function, G is the growth rate function. The spatial velocity is defined as the rate of change of position on the y-axis with respect to time t. Different kernels $K(x,y)$ and functions G and Z are considered. In particular, for the general homogenous kernel $K(\alpha x, \alpha y) = \alpha^\gamma K(x,y)$, and constant $G \neq 0$ and[3] $Z = 0$, Eq. (13) admits the infinitesimal generators

$$X_1 = \frac{\partial}{\partial t}, \quad X_2 = x\frac{\partial}{\partial x} + t\frac{\partial}{\partial t} - (2+\gamma)f\frac{\partial}{\partial f}. \tag{14}$$

For the case of $k = 0, 1, 2$ the general solutions of the determining equations were derived by the method of expanding coefficients of an admitted generator in Taylor series. The method of preliminary group classification also was applied extending Eq. (13) by a nonhomogeneous source (sink) term.

4.3 *Viscoelastic Materials with Memory*

The research in [12–14] deals with a linear viscoelastic models of homogeneous, aging materials with memory.[4]

In [12], the considered model describes the stress relaxation behavior of one-dimensional viscoelastic materials (allowing aging). The system consists of two partial differential equations and an integral equation:

$$v_t = \sigma_x, \; e_t = v_x, \; \varphi(\sigma) = e + \int_0^t H(t,\tau)e(\tau)\,d\tau, \; \varphi'(\sigma) \neq 0. \tag{15}$$

[3]Notice that if Z is constant, then using an equivalence transform it can be reduced to zero.

[4]Group analysis also was applied to other viscoelastic and nonlocal elastic models [20–22].

In this system time t and reference position x are independent variables, while the stress σ, the velocity v, and the strain e are dependent variables, $H(t,\tau)$ is the kernel of relaxation, and $\varphi(\sigma)$ is a smooth function of the stress.

Using the method of splitting applied in [11], the determining equations defining the admitted Lie group of system of Eq. (15) are solved. The solution gave us a complete group classification of Eq. (15) with respect to the function $\varphi(\sigma)$ and the kernel $H(t,\tau)$. The group classification separates all models into three classes: (a) the linear function $\varphi(\sigma) = E\sigma$; (b) the function $\varphi(\sigma) = \alpha\exp(\gamma\sigma) + \beta$, $(\alpha\gamma \neq 0)$; (c) the function $\varphi(\sigma) = \alpha\sigma^\beta + \gamma$, $(\alpha\beta(\beta - 1) \neq 0)$. Along with the group classification, representations of all invariant solutions and reduced equations are constructed in [12].

The results obtained in [13, 14] extend the study of [12] to the motions of thermoviscoelastic materials

$$c\sigma_x = v_t, \quad e_t = v_x, \quad \theta_{xx} = w_t,$$
$$\sigma = Ee + \int_0^t G(t,s)e(s)ds - \theta - \int_0^t L(t,s)\theta(s)ds, \tag{16}$$
$$w = e + \int_0^t L(t,s)e(s)ds + \theta + \int_0^t c(t,s)\theta(s)ds.$$

Here t, x are the independent variables, the σ, v, e, θ and w are the dependent variables, while E is constant, $G(t,s)$, $L(t,s)$ and $c(t,s)$ are relaxation functions.

Complete group classification of Eq. (16) is given. The study is separated into four different cases. It is shown that in each case, the general solution of the determining Eq. (16) corresponds to the Lie algebra with the generators

$$\begin{gathered} cX_1 = \partial_x, \quad X_2 = v\partial_v + \sigma\partial_\sigma + e\partial_e + \theta\partial_\theta + w\partial_w, \\ X_\alpha = \lambda_{tx}\partial_v + \lambda_{tt}\partial_\sigma + \lambda_{xx}\partial_e + \mu_t\partial_\theta + \mu_{xx}\partial_w, \end{gathered} \tag{17}$$

where $\lambda(t,x)$, $\mu(t,x)$ are solutions of the system

$$\begin{gathered} c\lambda_{tt} - E\lambda_{xx} + \mu_t - \int_0^t G(t,s)\lambda_{xx}(s)ds + \int_0^t L(t,s)\mu_t(s)ds = 0, \\ \mu_{xx} - \lambda_{xx} - \mu_t - \int_0^t L(t,s)\lambda_{xx}(s)ds - \int_0^t c(t,s)\mu_t(s)ds = 0. \end{gathered} \tag{18}$$

Notice that $\lambda(t,x) = tx$, $\mu(t,x) = 0$ is the trivial solution of system (18), and determines the generator

$$X_3 = \partial_v.$$

Using the two subalgebras $\{X_1, X_2\}$ and $\{X_1, X_2, X_3\}$, two classes of partially invariant solutions of Eq. (16) were studied.

4.4 Evolutionary Integro-Differential Equations Describing Nonlinear Waves

One of the most general evolution equations used in nonlinear wave physics is [23, 24]:

$$
\begin{aligned}
&(u_x - uu_t - w_{tt})_t = u_{yy} + u_{zz}, \\
&w = \int_0^\infty K(s)\, u(t-s)\, ds.
\end{aligned}
\tag{19}
$$

Here the variable t is the time, and x, y, z are the spatial Cartesian coordinates. The coordinate x is distinguished as a "longitudinal" one. It coincides with a preferred orientation of the wave propagation. Other coordinates y, z are identified as "transversal" ones.

The paper [25] provides a first step in application of the Lie group analysis to Eq. (19). The analysis of the determining equation for the integro-differential equation allows, in particular, to single out a class of kernels used for deriving mathematical models in medical applications of ultrasound [26].

For particular kernels the integro-differential equation (19) becomes a partial differential equation or a delay partial differential equation. In these cases the complete group classification of Eq. (19) was obtained. Complete study of particular cases is given in the paper. Along with admitted Lie groups, representations of exact solutions and reduced equations are constructed in the paper. Solutions and a physical interpretations of some of them are presented in [25].

4.5 Kinetic Equation in a Nonlinear Thermal Transport Problem

In [27] an application of group analysis for finding and classifying analytic solutions of the electron kinetic equations in a nonlinear thermal transport problem is discussed. An electron kinetic model is formulated by keeping only the first two harmonics in the expansion of the electron distribution function, $f_e = f_0 + \mu f_1$, where μ is the cosine of the angle between the electron velocity vector and the plasma inhomogeneity direction (x direction). The following set of kinetic equations for f_0 and f_1 corresponds to the diffusive approximation in a one-dimensional inhomogeneous plasma with immobile ions:

$$
\begin{aligned}
c3 f_{0t} + v f_{1x} - \mathscr{E} v^{-2}(v^2 f_1)_v - 3C_{ee}(f_0, f_0) = 0\,, \\
v f_{0x} - \mathscr{E} f_{0v} + \nu_{ei}(v) f_1 = 0\,,
\end{aligned}
\tag{20}
$$

where $\mathscr{E} = eE/m$, E is the ambipolar electric field defined by the quasineutrality condition for the zero electric current,

$$j \equiv \int_0^\infty v^6 \left(\mathscr{E} f_{0v} - v f_{0x}\right) dv = 0\,, \tag{21}$$

and $\nu_{ei}(v) = 4\pi e^4 Z n \Lambda / m^2 v^3 \equiv ZY/v^3$ is the velocity-dependent electron–ion collision frequency. The electron–electron collision term is a nonlinear integro-differential operator,

$$C_{ee} = \nu_{ee}(v) v \partial_v \left(f_0 I_0^0 + \frac{v}{3}\left(I_2^0 + J_{-1}^0\right) f_{0v} \right), \tag{22}$$

involving the velocity-dependent electron–electron collision frequency $\nu_{ee}(v) = Y/v^3$ and the first three Rosenbluth potentials

$$I_0^0 = \frac{4\pi}{n}\int_0^v v^2 f_0\, dv\,, \qquad I_2^0 = \frac{4\pi}{nv^2}\int_0^v v^4 f_0\, dv\,, \qquad J_{-1}^0 = \frac{4\pi}{n} v \int_v^\infty v f_0\, dv\,. \tag{23}$$

It is more convenient to use differential consequences of the latter Eq. (23),

$$I_{0v}^0 = \frac{4\pi}{n} v^2 f_0\,, \qquad I_{2v}^0 + \frac{2}{v} I_2^0 = \frac{4\pi}{n} v^2 f_0\,, \qquad J_{-1v}^0 - \frac{1}{v} J_{-1}^0 = -\frac{4\pi}{n} v^2 f_0\,. \tag{24}$$

In (20) and (22), the electron–electron and electron–ion collision frequencies $\nu_{ee}(v)$ and $\nu_{ei}(v)$ depend on v and n, the second-order momentum of the distribution function,

$$\nu_{ee}(v) = knv^{-3}\,, \qquad \nu_{ei}(v) = k_1 n v^{-3}\,, \qquad n = 4\pi \int_0^\infty v^2 f_0\, dv\,, \tag{25}$$

where $k = 4\pi e^4 \Lambda / m^2$ and $k_1 = kZ$.

The key idea of solving the determining equations was to find the symmetry group for differential equations (20) and (24) supplemented with the differential equalities

$$\mathscr{E}_v = 0\,, \qquad n_v = 0\,, \tag{26}$$

which are obvious from the physical standpoint. It gave five infinitesimal group generators of the so-called intermediate symmetry:

$$\begin{gathered} cX_1 = x\partial_x + v\partial_v + \mathscr{E}\partial_{\mathscr{E}} + 3n\partial_n, \\ X_2 = t\partial_t + x\partial_x - f_0\partial_{f_0} - f_1\partial_{f_1} - \mathscr{E}\partial_{\mathscr{E}} - n\partial_n\,, X_3 = \partial_t, \quad X_4 = \partial_x, \\ X_5 = x\partial_x + f_1\partial_{f_1} - \mathscr{E}\partial_{\mathscr{E}} - 2n\partial_n + 2I_0^0\partial_{I_0^0} + 2I_2^0\partial_{I_2^0} + 2J_{-1}^0\partial_{J_{-1}^0}\,. \end{gathered} \tag{27}$$

Verifying the invariance conditions for nonlocal relations (21) and (25) under the group transformations given by (27) yielded additional limitations on this group which excluded the generator X_5. Finally, admitted symmetries are defined by the generators X_1, X_2, X_3, and X_4. The optimal system of one and two-dimensional subalgebras was constructed, and representations of invariant solutions are presented in [27].

5 Symmetries of Delay Differential Equations

Our recent applications of the group analysis method to delay differential equations can be separated in two groups. Results of the first group [15, 28, 29] were obtained, where a delay considered unchangeable by admitted Lie group. Whereas in the studies presented in the second group [30, 31] the delay is allowed to be changed under the Lie group of transformations.

5.1 Nonlinear Klein–Gordon Equation

The delay in the delay differential equations in all previous applications [2] of the group analysis method to delay differential equations has been assumed to be constant. In [28, 29], the group analysis of the nonlinear Klein–Gordon equation with a time-varying delay was studied. The derived analysis is applied to the nonlinear two-dimensional equation

$$u_{tt} = u_{xx} + u_{yy} + g(u, \bar{u}), \; g_{\bar{u}}(u, \bar{u}) \neq 0, \tag{28}$$

where $\bar{u}(t, x, y) = u(t - \tau(t), x, y)$, $\tau(t)$ depends on t, and $\tau(t) > 0$.

First, the determining equations for equations with a time-varying delay are derived. Then the developed analysis is applied to Eq. (28). The complete group classification of this equation with respect to the arbitrary function g is obtained. All admitted Lie algebras are classified. These classifications are used for deriving invariant solutions. Representations of all invariant solutions are also given in [29].

5.2 Delay Ordinary Differential Equations

For ordinary differential equations with a single dependent variable, group classification (in the general case) is obtained by using the realizations of all Lie algebras nonequivalent with respect to a change of the variables. Lie gave the classification of all dissimilar Lie algebras (under complex change of variables) in two complex variables. The authors of [32] ordered the Lie classification and extended it to the real case [33]. Using differential invariants of these Lie algebras up to a studied order, representations of ordinary differential equations were found.

Similar strategy was applied in [34] for classification of second-order delay ordinary differential equations of the form

$$y'' = f(x, y, y_\tau, y', y'_\tau) \tag{29}$$

where $\tau > 0$ is a constant delay, $y_\tau = y(x-\tau)$ and $y'_\tau = y'(x-\tau)$. As a nonsingular change of the dependent and independent variables is not an equivalence transformation in contrast to ordinary differential equations, for delay differential equations the authors of [34] applied additional two steps. The following strategy was used:

(a) change the variables x and y

$$\bar{x} = h(x, y), \qquad \bar{y} = g(x, y); \tag{30}$$

(b) find invariants of the Lie algebra in the space of the changed variables $(\bar{x}, \bar{y}, \bar{y}_\tau, \bar{y}', \bar{y}'_\tau, \bar{y}'')$;
(c) use the found invariants to form a second-order delay ordinary differential equation.

The delay τ was not allowed being changed. Applying this strategy, representations of all second-order delay ordinary differential equations admitting a finite dimensional Lie algebra were obtained. It should be noted here that the application of steps (a) and (c) is a weakness of the used strategy. This weakness does not allow to direct use of the classification obtained in [34].

Recently [30, 31] a new strategy for classification of delay differential equations was applied. The strategy consists of use of differential invariants of the Lie algebras [33], and allowing for the delay to be changed. Using differential invariants of these Lie algebras up to a first order, complete group classification of first-order delay ordinary differential equations supplemented by equations for the delay were given in [30, 31].

6 Applications to Stochastic Differential Equations

Additional complications in applications of the group analysis method to stochastic differential equations are due to (a) the fact that the derivative of a composition of functions is calculated using the Itô formula, unlike in the case of deterministic equations; (b) stochastic differential equations contain integrals of two types (Itô and Riemann).[5]

[5]The determining equations are derived by equating the integrands of each of the integrals. It requires justification. Necessary and sufficient conditions for this justification were obtained recently in [35].

6.1 Symmetries of Stochastic Fluid Dynamics Equations

6.1.1 Determining Equations

On a complete probability space $(\Omega, \mathscr{F}, P)$ consider the stochastic Cauchy problem

$$v^i(t,y) = v^i(t_0,y) + \int_{t_0}^{t} A^i(\tau,y)\,d\tau + \sum_{j=1}^{m}\int_{t_0}^{t} B^{ij}(\tau,y)\,dw^j(\tau), \quad (i=\overline{1,n}) \tag{31}$$

where

$$A^i(t,y) = A^i(t,y,v(t,y),v_{y_k}(t,y),v_{y_k y_l}(t,y)), \quad B^{ij}(t,y) = B^{ij}(t,y,v(t,y)),$$

$w(t)$ is a vector of m independent standard Wiener processes, $v(t_0,y) = v_0(y)$ a vector of random functions, $t \in [t_0,T]$ and $y = (y_1,\ldots,y_N)^T \in R^N$. It is assumed that the set of Lie group of transformations belongs to the class of the change of the dependent and independent variables

$$r = \alpha(t), \quad y = h(t,x), \quad v = g(t,x,u), \tag{32}$$

where the functions $\alpha(t)$, $h(t,x)$ and $g(t,x,u)$ are locally invertible with respect to t, x and u.

The determining equations defining an admitted Lie group have the form

$$\widetilde{X}(A^i - u^i_t) = 0, \tag{33}$$

$$\widetilde{X}B^{ij} + \frac{1}{2}B^{ij}\psi_t - \sum_{k=1}^{n} B^{kj}\zeta^{u^i}_{u^k} = 0 \tag{34}$$

where

$$X = \psi(t)\partial_t + \xi^{x_k}(t,x)\partial_{x_k} + \zeta^{u^i}(t,x,u)\partial_{u^i},$$

is the infinitesimal generator of the group. The coefficients of the prolonged operator $\widetilde{X}$ of the generator X are the same as for differential equations except the coefficients $\zeta^{u^i_t}$ which are related with the derivatives u^i_t:

$$\zeta^{u^i_t} = \sum_{j=1}^{n}\left(A^j\psi_t + A^j\zeta^i_{u^j} + \frac{1}{2}\sum_{k=1}^{n}\left(\sum_{\sigma=1}^{m} B^{j\sigma}B^{k\sigma}\right)\zeta^i_{u^j u^k}\right).$$

6.1.2 The Gas Dynamics Equations

Group properties of stochastic gas- and hydro-dynamics equations were analyzed in [36].

Consider the gas dynamics partial differential equations with stochastic part,

$$\begin{aligned} &\rho_t + u\rho_x + \rho u_x = 0, \\ &p_t + up_x + \gamma p u_x = 0, \\ &du = -\left(uu_x + \tfrac{1}{\rho} p_x\right) dt + B(t, x, u, p, \rho)\, dW(t). \end{aligned} \tag{35}$$

The stochastic part, determined by the function $B(t, x, u, p, \rho)$, can be interpreted as a stochastic external force. This function presents an arbitrary element for the group classification of Eq. (35).

Applying the operator $\widetilde{X}$ to Eq. (35) and substituting $\rho_t = -(u\rho_x + \rho u_x)$, $p_t = -(up_x + \gamma p u_x)$, one obtains the determining equations:

$$\begin{aligned} &\widetilde{X}(\rho_t + u\rho_x + \rho u_x) = 0, \\ &\widetilde{X}(p_t + up_x + \gamma p u_x) = 0, \\ &\widetilde{X}\left[\left(-uu_x - \tfrac{1}{\rho} p_x\right) - u_t\right] = 0, \end{aligned} \tag{36}$$

$$\widetilde{X}B + \frac{B}{2}\psi_t - B\zeta^u_u = 0. \tag{37}$$

For the group classification of Eq. (35), one can apply the algebraic approach used in [37]. This approach is performed in two steps. First, one finds the general solution of the determining Eq. (36). For $\gamma \neq 3$ this solution gives that

$$X = c_1X_1 + c_2X_2 + c_3X_3 + c_4X_4 + c_5X_5 + c_6X_6$$

where c_i, $(i = 1, 2, \ldots, 6)$ are arbitrary constants, and

$$\begin{aligned} &X_1 = x\partial_x + u\partial_u - 2\rho\partial_\rho, \quad X_2 = \partial_t, \quad X_3 = t\partial_t - u\partial_u + 2\rho\partial_\rho, \\ &X_4 = \partial_x, \quad X_5 = t\partial_x + \partial_u, \quad X_6 = p\partial_p + \rho\partial_\rho. \end{aligned} \tag{38}$$

The generators X_i, $(i = 1, 2, \ldots, 6)$ compose a six-dimensional Lie algebra L_6. On the second step, using a subalgebra of the Lie algebra L_6 which provides the constants c_i, one solves Eq. (37).

Similar studies were given for the two-dimensional Navier Stokes stochastic partial differential equations,

$$\begin{aligned} du^1 &= \left[u^1_{x_1x_1} + u^1_{x_2x_2} - (u^1u^1_{x_1} + u^2u^1_{x_2} + p_{x_1})\right] dt + B^{11}dW_1(t) + B^{12}dW^2(t), \\ du^2 &= \left[u^2_{x_1x_1} + u^2_{x_2x_2} - (u^1u^2_{x_1} + u^2u^2_{x_2} + p_{x_2})\right] dt + B^{21}dW^1(t) + B^{22}dW^2(t), \\ &u^1_{x_1} + u^2_{x_2} = 0. \end{aligned} \tag{39}$$

where $B^{ij} = B^{ij}(t, x, u, p)$, $1 \le i, j \le 2$.

These studies demonstrated a first experience in application of the group analysis method for constructing invariant solutions of stochastic differential equations of the gas and hydro dynamics.

6.2 Trajectory Approach

In [38] a new approach for application of the group analysis method to one-dimensional stochastic ordinary differential equations was proposed. Using this approach, the problem of group analysis of stochastic differential equations reduces to the same problem for an ordinary differential equation whose right-hand side generally depends on the path of the Wiener process. The reduction to the analysis of ordinary differential equation is based on the following result derived by F.S.Nasyrov in [35].

Let on the probability space $(\Omega, F, (F_t)_{0\le t\le T}, P)$ be given a Brownian motion $W(t)$, $t \in [0, T]$. Consider the stochastic differential equation

$$u(t) = u_0 + \int_0^t b(s, u(s))ds + \int_0^t \sigma(s, u(s)) * dW(s), \ t \in [0, T], \tag{40}$$

where the second integral on the right-hand is a stochastic Stratonovich integral. We assume that the coefficients $b(s, \phi)$ and $\sigma(s, \phi)$ satisfy the following conditions (C and D are constant):

(1) the condition of linear growth: $|b(t, \phi)| + |\sigma(t, \phi)| \le C(1 + |\phi|)$;
(2) the Lipschitz condition: $|b(t, x) - b(t, y)| + |\sigma(t, x) - \sigma(t, y)| \le D|x - y|$;
(3) $\sigma^2(t, \phi) > 0, \forall t \in [0, T]$.

These assumptions ensure the existence and uniqueness of the Cauchy problem of Eq. (40). Furthermore, condition (3) can be used to determine for every t the function $\phi^*(t, v)$ which is inverse to the function

$$v = \gamma(t, \Phi) = \int \frac{d\Phi}{\sigma(t, \Phi)}.$$

It has been proved in [35] that a solution $u(t)$ of Eq. (40) has the structure:

$$u(t) = \phi^*(t, W(t) + C(t)), \tag{41}$$

where the deterministic function $\phi^*(t, v)$ was defined above, and the smooth function $C(t)$ is the solution of the Cauchy problem

$$C'(t) = H(t, W(t) + C(t)), \qquad \phi^*(0, W(0) + C(0)) = u_0, \tag{42}$$

and

$$H(t, v) = \frac{b(t, \phi^*(t, v)) - (\phi^*)_t(t, v)}{\sigma(t, \phi^*(t, v))}.$$

Thus, one considers the Lie group admitted by Eq. (42). These transformations have the form $\bar{t} = f(t, a)$, $\bar{C} = h(t, C, a)$ under which Eq. (42) becomes unchangeable:

$$\bar{C}'(\bar{t}) = H(\bar{t}, \bar{C}(\bar{t}) + \bar{W}(\bar{t})), \tag{43}$$

where

$$\bar{C}(\bar{t}) = h(f(\bar{t}, -a), C(f(\bar{t}, -a)), a). \tag{44}$$

6.3 Linearization of Systems of Two Second-Order Equations

Linear stochastic ordinary differential equations play a role similar to that of linear equations in the deterministic theory of ordinary differential equations. However, the change of variables in stochastic ordinary differential equations differs from that in ordinary differential equations due to the Itô formula. The transformation of nonlinear stochastic ordinary differential equations into linear ones via an invertible stochastic mapping proves to be useful in obtaining the closed form solutions.

Consider the system of two second-order SODEs,

$$\begin{aligned} d\dot{X} &= f_1(t, X, Y, \dot{X}, \dot{Y})\,dt + g_1(t, X, Y, \dot{X}, \dot{Y})\,dW \\ d\dot{Y} &= f_2(t, X, Y, \dot{X}, \dot{Y})\,dt + g_2(t, X, Y, \dot{X}, \dot{Y})\,dW, \end{aligned} \tag{45}$$

where f_i and g_i, $(i = 1, 2)$ are deterministic functions and dW is the infinitesimal increment of the Wiener process. System (45) is said to be linear if the functions f_i and g_i are linear functions with respect to variables X and Y and their respective derivatives. For the linearization problem one considers the class of equations equivalent to linear equations with respect to the change of variables

$$x_1 = \varphi(t, x, y), \quad p_1 = \varphi_2(t, x, y, p, q); \quad y_1 = \psi(t, x, y), \quad q_1 = \psi_2(t, x, y, p, q) \tag{46}$$

with

$$\Delta = \varphi_x \psi_y - \varphi_y \psi_x \neq 0,$$

In [39], the complete solution of the linearization problem of systems of two second-order stochastic ordinary differential equations is presented. Necessary and sufficient conditions for linearization by an invertible transformation are given in terms of coefficients of the system. Some illustrative examples are provided. Moreover, a code using REDUCE for checking whether a system of two second-order stochastic ordinary differential equations is linearizable has developed.

Acknowledgements This research was supported by Russian Science Foundation Grant No 18-11-00238 'Hydrodynamics-type equations: symmetries, conservation laws, invariant difference schemes'.

References

1. Meleshko, S.V.: Methods for Constructing Exact Solutions of Partial Differential Equations. Mathematical and Analytical Techniques with Applications to Engineering. Springer, New York (2005)
2. Grigoriev, YuN, Ibragimov, N.H., Kovalev, V.F., Meleshko, S.V.: Symmetries of Integro-Differential Equations and their Applications in Mechanics and Plasma Physics. Lecture Notes in Physics, vol. 806. Springer, Berlin (2010)
3. Grigoriev, Yu.N., Meleshko, S.V.: Investigation of invariant solutions of the Boltzmann kinetic equation and its models (1986). Preprint of Institute of Theoretical and Applied Mechanics
4. Bobylev, A.V., Ibragimov, N.H.: Relationships between the symmetry properties of the equations of gas kinetics and hydrodynamics. J. Math. Model. **1**(3), 100–109 (1989)
5. Taranov, V.B.: On symmetry of one-dimensional high frequency motion of noncollision plasma. J. Tech. Phys. **46**(6), 1271–1277 (1976)
6. Bunimovich, A.I., Krasnoslobodtsev, A.V.: Invariant-group solutions of kinetic equations. Mechan. Jzydkosti i gasa **4**, 135–140 (1982)
7. Bunimovich, A.I., Krasnoslobodtsev, A.V.: On some invariant transformations of kinetic equations. Vestn. Mosc. State Univ. Ser. 1. Mat. Mech. **4**, 69–72 (1983)
8. Grigoriev, Yu.N., Meleshko, S.V.: Group analysis of the integro-differential Boltzman equation. Dokl. AS USSR **297**(2), 323–327 (1987)
9. Lin, F., Flood, A.E., Meleshko, S.V.: Exact solutions of population balance equations for aggregation, breakage and growth processes. Appl. Math. Comput. **307**(15), 193–203 (2017)
10. Lin, F., Flood, A.E., Meleshko, S.V.: Exact solutions of the population balance equation involving aggregation, breakage and growth processes and particle transport in one dimension. Commun. Nonlinear Sci. Numer. Simul. **59**, 255–271 (2018)
11. Meleshko, S.V.: On group properties of viscoelastic media motion equations. Modelirovanie v mechanike **2**(19)(4), 116–126 (1988)
12. Zhou, L.Q., Meleshko, S.V.: Group analysis of integro-differential equations describing stress relaxation behavior of one-dimensional viscoelastic materials. Int. J. Non-Linear Mech. **77**, 223–231 (2015)
13. Zhou, L.Q., Meleshko, S.V.: Invariant and partially invariant solutions for a linear thermoviscoelasticity. Contin. Mech. Thermodyn. **29**, 207–224 (2017)
14. Zhou, L.Q., Meleshko, S.V.: Symmetry groups of integro-differential equations for linear thermoviscoelastic materials with memory. J. Appl. Mech. Tech. Phys. **58**(4), 22–45 (2017)
15. Long, F.-S., Karnbanjong, A., Suriyawichitseranee, A., Grigoriev, Y.N., Meleshko, S.V.: Application of a Lie group admitted by a homogeneous equation for group classification of a corresponding inhomogeneous equation. Commun. Nonlinear Sci. Numer. Simul. **48**, 350–360 (2017)
16. Bobylev, A.V.: On exact solutions of the Boltzmann equation. Dokl. AS USSR **225**(6), 1296–1299 (1975)
17. Karnbanjong, A., Suriyawichitseranee, A., Grigoriev, Yu.N., Meleshko, S.V.: Preliminary group classification of the full Boltzmann equation with a source term. AIP Conf. Proc. (030062) (2017)
18. Meleshko, S.V., Grigoriev, Yu.N., Karnbanjong, A., Suriyawichitseranee, A.: Invariant solutions in explicit form of the Boltzmann equation with a source term. J. Phys.: Conf. Ser. (012063) (2017)

19. Lin, F., Flood, A.E., Meleshko, S.V.: Exact solutions of population balance equation. Commun. Nonlinear Sci. Numer. Simul. **36**, 378–390 (2016)
20. Senashov, S.I.: Group classification of viscoelastic bar equation. Modelirovanie v mechanike **4**(21)(1), 69–72 (1990)
21. Özer, T.: On the symmetry group properties of equations of nonlocal elasticity. Mech. Res. Commun. **26**(6), 725–733 (1999)
22. Özer, T.: Symmetry group classification for two-dimensional elastodynamics problems in nonlocal elasticity. Int. J. Eng. Sci. **44**(18), 2193–2211 (2003)
23. Rudenko, O.V., Soluyan, S.I.: Theoretical Foundations of Nonlinear Acoustics. Plenum Consultants Bureau, New York (1977)
24. Rudenko, O.V., Soluyan, S.I., Khokhlov, R.V.: Problems of the theory of nonlinear acoustics. Sov. Phys. Acoust. **20**(3), 356–359 (1974)
25. Ibragimov, N.H., Meleshko, S.V., Rudenko, O.V.: Group analysis of evolutionary integro-differential equations describing nonlinear waves: general model. J. Phys. A: Math. Theor. **44**(315201) (2011)
26. Sarvazyan, A.P., Rudenko, O.V., Swanson, S.D., Fowlkes, J.B., Emelianov, S.Y.: Shear wave elasticity imaging: a new ultrasonic technology of medical diagnostics. Ultrasound Med. Biol. **24**(9), 1419–1435 (1998)
27. Ibragimov, N.H., Kovalev, V.F., Meleshko, S.V., Bychenkov, V.Yu.: Group analysis of kinetic equations in a nonlinear thermal transport problem. Int. J. Non-Linear Mech. **71**, 1–7 (2015)
28. Long, F.-S., Meleshko, S.V.: On the complete group classification of the one-dimensional nonlinear Klein-Gordon equation with a delay. Math. Methods Appl. Sci. **39**(12), 3255–3270 (2016)
29. Long, F.-S., Meleshko, S.V.: Symmetry analysis of the nonlinear two-dimensional Klein-Gordon equation with a time-varying delay. Math. Methods Appl. Sci. **40**(13), 4658–4673 (2017)
30. Dorodnitsyn, V.A., Kozlov, R., Meleshko, S.V., Winternitz, P.: Lie group classification of first-order delay ordinary differential equations. J. Phys. A: Math. Theor. (in press)
31. Dorodnitsyn, V.A., Kozlov, R., Meleshko, S.V., Winternitz, P.: Linear or linearizable first-order delay ordinary differential equations and their lie point symmetries. J. Phys. A: Math. Theor. (in press)
32. Gonzalez-Lopez, A., Kamran, N., Olver, P.J.: Lie algebras of differential operators in two complex variables. Am. J. Math. **114**, 1163–1185 (1992)
33. Gonzalez-Lopez, A., Kamran, N., Olver, P.J.: Lie algebras of vector fields in the real plane. Proc. Lond. Math. Soc. **64**, 339–368 (1992)
34. Pue-on, P., Meleshko, S.V.: Group classification of second-order delay ordinary differential equation. Commun. Nonlinear Sci. Numer. Simul. **15**, 1444–1453 (2010). https://doi.org/10.1016/j.cnsns.2009.06.013
35. Nasyrov, F.S.: Local Times, Symmetric Integrals and Stochastic Analysis. Fizmatlit, Moscow (2011)
36. Meleshko, S.V., Sumrum, O., Schulz, E.: Application of group analysis to stochastic equations of fluid dynamics. J. Appl. Mech. Tech. Phys. **54**(1), 21–33 (2013)
37. Grigoriev, Yu.N., Meleshko, S.V., Suriyawichitseranee, A.: On group classification of the spatially homogeneous and isotropic Boltzmann equation with sources ii. Int. J. Non-Linear Mech. **61**, 15–18 (2014)
38. Nasyrov, F.S., Abdullin, M.A., Meleshko, S.V.: On new approach to group analysis of one-dimensional stochastic differential equations. J. Appl. Mech. Tech. Phys. **55**(2), 1–9 (2014)
39. Mkhize, T.G., Govinder, K., Moyo, S., Meleshko, S.V.: Linearization criteria for systems of two second-order stochastic ordinary differential equations. Appl. Math. Comput. **301**, 25–35 (2017)

A Note on the Multiplier Approach for Derivation of Conservation Laws for Partial Differential Equations in the Complex Domain

R. Naz and F. M. Mahomed

Abstract We study the conservation laws of scalar partial differential equations (PDEs) with two real independent variables in the complex plane. The complex PDE is split into a system of two real coupled or uncoupled PDEs. We invoke the multiplier method for the derivation of conserved quantities for the complex PDEs and their split systems. The approach is applied to both variational and non-variational complex PDEs. The decomposed complex multipliers of the complex PDE yields two real multipliers for the related split system in the real plane. The multipliers of the split system are derived by utilizing the multiplier method. The multipliers of the split system are compared with the multipliers of the complex PDE after decomposition of the complex multipliers. It is demonstrated that the split multipliers of the complex PDE are not in general the same as the multipliers of the decomposed system of real PDEs. They are shown to be identical when all the multipliers of the complex PDE have either pure real or imaginary parts. We moreover look at the number of conserved vectors that arise by a complex split and from the real system by use of the multipliers.

Keywords Multiplier approach · Complex domain · Conservation laws

1 Introduction

Conserved quantities are of significance in the study of differential equations and their applications. There are different approaches in the construction of conservation laws and these are discussed in [1]. The direct construction method as presented in

R. Naz (✉)
Lahore School of Economics, Centre for Mathematics and Statistical Sciences, Lahore 53200, Pakistan
e-mail: drrehana@lahoreschool.edu.pk

F. M. Mahomed
School of Computer Science and Applied Mathematics, DST-NRF Centre of Excellence in Mathematical and Statistical Sciences, University of the Witwatersrand, Johannesburg, Wits 2050, South Africa

V. G. Kac et al. (eds.), *Symmetries, Differential Equations and Applications*, Springer Proceedings in Mathematics & Statistics 266,
https://doi.org/10.1007/978-3-030-01376-9_7

[2, 3] can be invoked for the determination of conserved quantities for both variational and non-variational equations. Complex symmetry analysis for ordinary differential equations (ODEs) in the complex domain are investigated in [4–6]. Recently the idea of hypercomplex analysis of ODEs was initiated in [7]. The symmetries for complex partial differential equations (PDEs) and their decomposed systems are studied in [8]. It was shown that the split Lie-like operators of the complex PDEs are in general not symmetries of the decomposed system of real PDEs. Also a proposition provides for when the Lie-like operators are indeed Lie symmetries of the decomposed system [8]. A complex variational method for the variational PDEs was developed by Naz and Mahomed [9]. It was shown that the decomposed conserved vectors of the complex PDE were identical to the conserved vectors of the decomposed system of real PDEs for coupled systems whereas these were different for the uncoupled split system [9].

Here we study complex PDEs with two real independent variables which are variational and non-variational. The complex PDE is split into a system of two real PDEs. The multiplier approach is utilized for the derivation of the conservation laws for the complex PDEs and their split systems. The split of the complex multipliers of a complex PDE yields two real multipliers for the associated decomposed system in the real plane. The multipliers of the split system are derived by utilizing the multiplier approach as well. The multipliers of the split system are checked against the multipliers after split of the complex multipliers of the complex PDE via examples. It is shown that the decomposed multipliers of the complex PDE are not in general identical with the multipliers of the decomposed system of real PDEs. We show them to be identical when all the multipliers of the complex PDE have either pure real or imaginary parts.

The paper is organized in the following manner: in Sect. 2, we provide the multiplier approach for the complex PDEs and their decomposed systems. In Sect. 3, we establish the conservation laws for several complex PDEs and their decomposed systems. We investigate both Lagrangian and non-variational complex PDEs. In Sect. 4, the concluding remarks are presented.

2 The Multiplier Approach: Complex and Real Domains

Suppose we have the rth-order complex PDE

$$E\big(t, x,\ w,\ w_t,\ w_{(1)},\ w_{(2)}, \ldots, w_{(r)}\big) = 0\ , \tag{1}$$

where t is the time, x is a real independent variable, w is a complex-valued function of x and t with the coordinates $w = u + iv$ and $w_{(r)}$ being the rth-derivative of w with respect to x. Equation (1) decomposes into a system of two, real coupled or uncoupled, PDEs

$$E_1\big(t, x, u, v, u_t, v_t, u_{(1)}, v_{(1)}, u_{(2)}, v_{(2)}, \ldots, u_{(r)}, v_{(r)}\big) = 0,$$
$$E_2\big(t, x, u, v, u_t, v_t, u_{(1)}, v_{(1)}, u_{(2)}, v_{(2)}, \ldots, u_{(r)}, v_{(r)}\big) = 0, \qquad (2)$$

in which $E = E_1 + iE_2$.

1. A conserved vectors for (1) satisfies

$$D_t\mathbf{T}^1 + D_x\mathbf{T}^2 = 0. \qquad (3)$$

By setting $\mathbf{T^j} = R^j + iI^j$, we have that

$$D_t R^1 + D_x R^2 = 0$$
$$D_t I^1 + D_x I^2 = 0 \qquad (4)$$

and thus (R^1, R^2) and (I^1, I^2) are two conserved vectors for the decomposed system (2).

2. Suppose that Eq. (1) has a complex multiplier $\mathbf{\Lambda}$ depending upon the usual dependent, independent variables and derivatives of the dependent variables up to certain fixed order. The complex multiplier satisfies

$$\mathbf{\Lambda}E = D_t\mathbf{T}^1 + D_x\mathbf{T}^2 \qquad (5)$$

where $\mathbf{T}^1$ and $\mathbf{T}^2$ are conserved vectors of the complex PDE (1). Equation (5) splits into

$$\mathbf{\Lambda}^1 E_1 - \mathbf{\Lambda}^2 E_2 = D_t R^1 + D_x R^2 \qquad (6)$$

$$\mathbf{\Lambda}^1 E_2 + \mathbf{\Lambda}^2 E_1 = D_t I^1 + D_x I^2$$

where $\mathbf{\Lambda} = \mathbf{\Lambda}^1 + i\mathbf{\Lambda}^2$, $\mathbf{\Lambda}^1(t, x, u, v, u_t, v_t, u_x, v_x, \ldots)$ and $\mathbf{\Lambda}^2(t, x, u, v, u_t, v_t, u_x, v_x, \ldots)$.

3. Let (T^1, T^2) be a conserved vector for the decomposed system (2). For multipliers of the form $\Lambda^1(t, x, u, v, u_t, v_t, u_x, v_x, \ldots)$ and $\Lambda^2(t, x, u, v, u_t, v_t, u_x, v_x, \ldots)$, we have

$$\Lambda^1 E_1 + \Lambda^2 E_2 = D_t T^1 + D_x T^2. \qquad (7)$$

4. From Eqs. (6) and (7) we observe that if

$$\mathbf{\Lambda}^1 = \Lambda^1, \mathbf{\Lambda}^2 \mapsto -\Lambda^2 \qquad (8)$$

then the real part of the conserved vector for Eq. (1) is the conserved vector for the decomposed system (2). Similarly, Eqs. (6) and (7) yield

$$\mathbf{\Lambda}^2 = \Lambda^1, \mathbf{\Lambda}^1 = \Lambda^2 \qquad (9)$$

so that the imaginary part of the conserved vector for Eq. (1) is the conserved vector for the decomposed system (2).

3 Applications

In this section, we invoke the multiplier method for the derivation of conservation laws for several complex PDEs and their decomposed systems for both Lagrangians and non-variational problems. The applications include the nonlinear spherical KdV, the Maxwellian tails equation, the Boussinesq equation and the wave equation in the complex domain.

3.1 The Nonlinear Spherical KdV Equation in the Complex Domain

Consider the nonlinear spherical KdV equation in the complex domain

$$w_t + 6ww_x + w_{xxx} + \frac{w}{t} = 0, \tag{10}$$

where $w = u + iv$ and t is the time. It is a non-variational problem. The multipliers of the form $\mathbf{\Lambda}(t, x, w)$ for (10) are

$$\mathbf{\Lambda}_{(1)} = t,\ \ \mathbf{\Lambda}_{(2)} = t^2 w,\ \ \mathbf{\Lambda}_{(3)} = tx + 6t^2(1 - \ln t)w. \tag{11}$$

For the multipliers (11), we have the following conservation law fluxes:

$$\begin{aligned}
\mathbf{T}_1^1 = tw,\ \mathbf{T}_1^2 = 3tw^2 + tw_{xx},\\
\mathbf{T}_2^1 = \frac{1}{2}t^2w^2,\ \mathbf{T}_2^2 = 2t^2w^3 + t^2ww_{xx} - \frac{1}{2}t^2w_x^2,\\
\mathbf{T}_3^1 = tw\big[x + 3tw(1 - \ln t)\big],\\
\mathbf{T}_3^2 = t^2(1 - \ln t)(12w^3t + 6ww_{xx} - 3w_x^2) + 3w^2tx + txw_{xx} - tw_x.
\end{aligned} \tag{12}$$

If we consider higher-order multipliers we obtain the same conserved vectors as in (12).

Equation (10) splits into the following coupled system

$$\begin{aligned}
E_1 = u_t + 6uu_x - 6vv_x + u_{xxx} + \frac{u}{t} = 0,\\
E_2 = v_t + 6vu_x + 6uv_x + v_{xxx} + \frac{v}{t} = 0.
\end{aligned} \tag{13}$$

By setting $\mathbf{\Lambda}_{(j)} = \mathbf{\Lambda}^1_{(j)} + i\mathbf{\Lambda}^2_{(j)}$ in (11) yields the following split multipliers:

$$\begin{aligned}
\mathbf{\Lambda}^1_{(1)} &= t,\ \mathbf{\Lambda}^2_{(1)} = 0,\\
\mathbf{\Lambda}^1_{(2)} &= t^2 u,\ \mathbf{\Lambda}^2_{(2)} = t^2 v,\\
\mathbf{\Lambda}^1_{(3)} &= tx + 6t^2(1-\ln t)u,\ \mathbf{\Lambda}^2_{(3)} = 6t^2(1-\ln t)v.
\end{aligned} \tag{14}$$

By using $\mathbf{T^j} = R^j + iI^j$ in (12), we have

$$\begin{aligned}
R_1^1 &= tu,\ R_1^2 = t(3u^2 - 3v^2 + u_{xx}),\\
I_1^1 &= tv,\ I_1^2 = t(6uv + v_{xx}),\\
R_2^1 &= \frac{t^2}{2}\left(u^2 - v^2\right),\\
R_2^2 &= t^2\left(2u^3 - 6uv^2 - \frac{1}{2}u_x^2 + \frac{1}{2}v_x^2 + uu_{xx} - vv_{xx}\right),\\
I_2^1 &= t^2 uv,\ I_2^2 = t^2\left(6u^2 v - 2v^3 + uv_{xx} + vu_{xx} - u_x v_x\right),\\
R_3^1 &= 3t^2(1-\ln t)(u^2 - v^2) + txu,\\
R_3^2 &= t^2(1-\ln t)(12u^3 - 36uv^2 + 6uu_{xx} - 6vv_{xx} - 3u_x^2 + 3v_x^2)\\
&\quad +3txu^2 - 3txv^2 - tu_x + txu_{xx},\\
I_3^1 &= 6t^2(1-\ln t)uv + txv,\\
I_3^2 &= t^2(1-\ln t)(36u^2 v - 12v^3 + 6uv_{xx} + 6vu_{xx} - 6u_x v_x)\\
&\quad +6txuv - tv_x + txv_{xx}.
\end{aligned} \tag{15}$$

The multiplier approach on the decomposed system (13) gives the following six multipliers:

$$\begin{aligned}
\Lambda^1_{(1)} &= t,\ \Lambda^2_{(1)} = 0,\\
\Lambda^1_{(2)} &= 0,\ \Lambda^2_{(2)} = t,\\
\Lambda^1_{(3)} &= t^2 u,\ \Lambda^2_{(3)} = -t^2 v,\\
\Lambda^1_{(4)} &= t^2 v,\ \Lambda^2_{(4)} = t^2 u,\\
\Lambda^1_{(5)} &= 6ut^2\ln t - tx,\ \Lambda^2_{(5)} = -6vt^2\ln t,\\
\Lambda^1_{(6)} &= 6vt^2\ln t,\ \Lambda^2_{(6)} = 6ut^2\ln t - tx,
\end{aligned} \tag{16}$$

and the corresponding conserved vectors are

$$\begin{aligned}
T_1^1 &= tu,\ T_1^2 = t(3u^2 - 3v^2 + u_{xx}),\\
T_2^1 &= tv,\ T_2^2 = t(6uv + v_{xx}),\\
T_3^1 &= \frac{t^2}{2}\left(u^2 - v^2\right),
\end{aligned}$$

$$
\begin{aligned}
T_3^2 &= t^2\left(-6uv^2 + 2u^3 + uu_{xx} - \frac{1}{2}u_x^2 - vv_{xx} + \frac{1}{2}v_x^2\right),\\
T_4^1 &= t^2uv,\ T_4^2 = t^2\left(6u^2v - 2v^3 + uv_{xx} + vu_{xx} - u_xv_x\right),\\
T_5^1 &= 3t^2\ln t(u^2 - v^2) - txu,\\
T_5^2 &= t^2\ln(t)(12u^3 - 36uv^2 + 6uu_{xx} - 6vv_{xx} - 3u_x^2 + 3v_x^2)\\
&\quad -3xtu^2 + 3xtv^2 + tu_x - txu_{xx}\\
T_6^1 &= 6t^2uv\ln t - txv,\\
T_6^2 &= t^2\ln t(36u^2v - 12v^3 + 6uv_{xx} + 6vu_{xx} - 6u_xv_x)\\
&\quad -6txuv + tv_x - txv_{xx}.
\end{aligned}
\tag{17}
$$

The split multipliers are different from the multipliers of the decomposed system but they satisfy (8) or (9). It is of value to see that $R_3^i = 6T_3^i - T_5^i$ and $I_3^i = 6T_4^i - T_6^i$, where $i = 1, 2$. For the nonlinear spherical KdV equation, the split conversed vectors are indeed conserved vectors for the decomposed coupled system as they should be. We have used the GeM package [10, 11] for computation of the multipliers and conservation law fluxes.

3.2 The Complex Maxwellian Tails Equation

Next we consider a variational problem. We apply the multiplier method on the complex Maxwellian tails equation. It decomposes into a coupled system of two real PDEs. From Table 1, we observe that the split multipliers are different from the multipliers of the decomposed system. The split conversed vectors are conserved vectors for the decomposed coupled system.

3.3 The Complex Boussinesq Equation

We apply the multiplier method to the fourth-order Boussinesq equation in the complex domain. From Table 2, we observe that the multipliers are purely real. The multipliers of the decomposed system and the multipliers after split of the complex multipliers of the complex Boussinesq equation are the same as all the multipliers are purely real. Also, the split conserved vectors of the complex Boussinesq equation are identical in number to the conserved vectors of the decomposed system of the real coupled PDEs.

For the complex nonlinear spherical KdV, the complex Maxwellian tails equation and the complex Boussinesq equation the decomposed system of PDEs is coupled. We see that the split conserved vectors of the complex PDE are the same number as the conserved vectors of the decomposed system of real PDEs in the case of the coupled system for both Lagrangian and non-variational problems.

Table 1 Multipliers and conserved vectors of the nonlinear complex and split Maxwallian tail equation

The Maxwellian equation: $w_{tx} + w_x + w^2 = 0$

Complex multipliers and conserved vectors	Split multipliers and conserved vectors
$\mathbf{\Lambda}_{(1)} = e^{2t} w_x$ $\mathbf{T}_1^1 = \frac{1}{2} e^{2t} w_x^2 \; \mathbf{T}_1^2 = \frac{1}{3} e^{2t} w^3$	$\mathbf{\Lambda}_{(1)}^1 = e^{2t} u_x, \; \mathbf{\Lambda}_{(1)}^2 = e^{2t} v_x$ $R_1^1 = \frac{1}{2} e^{2t} (u_x^2 - v_x^2)$ $R_1^2 = \frac{1}{3} e^{2t} (u^3 - 3uv^2)$ $I_1^1 = e^{2t} u_x v_x$ $I_1^2 = \frac{1}{3} e^{2t} (3u^2 v - v^3)$
$\mathbf{\Lambda}_{(2)} = e^{2t} (w + w_t + x w_x)$ $\mathbf{T}_2^1 = e^{2t} (\frac{1}{3} w^3 + \frac{1}{2} w w_x + \frac{1}{2} x w_x^2)$ $\mathbf{T}_2^2 = e^{2t} (\frac{1}{3} x w^3 + \frac{1}{2} w w_t + \frac{1}{2} w_t^2)$	$\mathbf{\Lambda}_{(2)}^1 = e^{2t} (u + u_t + x u_x),$ $\mathbf{\Lambda}_{(2)}^2 = e^{2t} (v + v_t + x v_x)$ $R_2^1 = e^{2t} [\frac{1}{3} (u^3 - 3uv^2)$ $+ \frac{1}{2} u u_x - \frac{1}{2} v v_x + \frac{x}{2} (u_x^2 - v_x^2)]$ $R_2^2 = e^{2t} [\frac{x}{3} (u^3 - 3uv^2)$ $+ \frac{1}{2} u u_t - \frac{1}{2} v v_t + \frac{1}{2} u_t^2 - \frac{1}{2} v_t^2]$ $I_2^1 = e^{2t} [\frac{1}{3} (3u^2 v - v^3)$ $+ \frac{1}{2} v u_x + \frac{1}{2} u v_x + x u_x v_x]$ $I_2^2 = e^{2t} [\frac{x}{3} (3u^2 v - v^3)$ $+ \frac{1}{2} v u_t + \frac{1}{2} u v_t + u_t v_t]$
$\mathbf{\Lambda}_{(3)} = e^{3t} (w + w_t)$ $\mathbf{T}_3^1 = e^{3t} (\frac{1}{3} w^3 + \frac{1}{2} w w_x)$ $\mathbf{T}_3^2 = e^{3t} (-\frac{1}{4} w^2 + \frac{1}{2} w w_t + \frac{1}{2} w_t^2)$	$\mathbf{\Lambda}_{(5)}^1 = e^{3t} (u + u_t), \; \mathbf{\Lambda}_{(5)}^2 = e^{3t} (v + v_t)$ $R_3^1 = e^{3t} [\frac{1}{3} (u^3 - 3uv^2)$ $+ \frac{1}{2} u u_x - \frac{1}{2} v v_x]$ $R_3^2 = e^{3t} [-\frac{1}{4} (u^2 - v^2)$ $+ \frac{1}{2} u u_t - \frac{1}{2} v v_t + \frac{1}{2} u_t^2 - \frac{1}{2} v_t^2]$ $I_3^1 = e^{3t} [\frac{1}{3} (3u^2 v - v^3)$ $+ \frac{1}{2} v u_x + \frac{1}{2} u v_x]$ $I_3^2 = e^{3t} [-\frac{1}{2} u v$ $+ \frac{1}{2} v u_t + \frac{1}{2} u v_t + u_t v_t]$

$u_{tx} + u_x + u^2 - v^2 = 0, \; v_{tx} + v_x + 2uv = 0$

Multipliers for the split system	Conserved vectors for the split system
$\Lambda_{(1)}^1 = e^{2t} u_x, \; \Lambda_{(1)}^2 = -e^{2t} v_x$	$T_1^1 = \frac{1}{2} e^{2t} (u_x^2 - v_x^2)$ $T_1^2 = \frac{1}{3} e^{2t} (u^3 - 3uv^2)$
$\Lambda_{(2)}^1 = e^{2t} v_x, \; \Lambda_{(2)}^2 = e^{2t} u_x$	$T_2^1 = e^{2t} u_x v_x$ $T_2^2 = \frac{1}{3} e^{2t} (3u^2 v - v^3)$
$\Lambda_{(3)}^1 = e^{2t} (u + u_t + x u_x)$	$T_3^1 = e^{2t} [\frac{1}{3} (u^3 - 3uv^2)$ $+ \frac{1}{2} u u_x - \frac{1}{2} v v_x + \frac{x}{2} (u_x^2 - v_x^2)]$
$\Lambda_{(3)}^2 = -e^{2t} (v + v_t + x v_x)$	$T_3^2 = e^{2t} [\frac{x}{3} (u^3 - 3uv^2)$ $+ \frac{1}{2} u u_t - \frac{1}{2} v v_t + \frac{1}{2} u_t^2 - \frac{1}{2} v_t^2]$
$\Lambda_{(4)}^1 = e^{2t} (v + v_t + x v_x), \; \Lambda_{(4)}^2 =$ $e^{2t} (u + u_t + x u_x)$	$T_4^1 = e^{2t} [\frac{1}{3} (3u^2 v - v^3)$ $+ \frac{1}{2} v u_x + \frac{1}{2} u v_x + x u_x v_x]$ $T_4^2 = e^{2t} [\frac{x}{3} (3u^2 v - v^3)$ $+ \frac{1}{2} v u_t + \frac{1}{2} u v_t + u_t v_t]$

(continued)

Table 1 (continued)

The Maxwellian equation: $w_{tx} + w_x + w^2 = 0$	
Complex multipliers and conserved vectors	Split multipliers and conserved vectors
$\Lambda^1_{(5)} = e^{3t}(u + u_t),\ \Lambda^2_{(5)} = -e^{3t}(v + v_t)$	$T^1_5 = e^{3t}[\frac{1}{3}(u^3 - 3uv^2)$ $+\frac{1}{2}uu_x - \frac{1}{2}vv_x]$ $T^2_5 = e^{3t}[-\frac{1}{4}(u^2 - v^2)$ $+\frac{1}{2}uu_t - \frac{1}{2}vv_t + \frac{1}{2}u_t^2 - \frac{1}{2}v_t^2]$
$\Lambda^1_{(6)} = e^{3t}(v + v_t),\ \Lambda^2_{(6)} = e^{3t}(u + u_t)$	$T^1_6 = e^{3t}[\frac{1}{3}(3u^2v - v^3)$ $+\frac{1}{2}vu_x + \frac{1}{2}uv_x]$ $T^2_6 = e^{3t}[-\frac{1}{2}uv$ $+\frac{1}{2}vu_t + \frac{1}{2}uv_t + u_tv_t]$

Table 2 Multipliers and conserved vectors of the nonlinear complex and split Boussinesq equation

The Boussinesq equation: $w_{tt} - w_{xx} + 3ww_{xx} + 3w_x^2 + w_{xxxx} = 0$	
Complex multipliers and conserved vectors	Split multipliers and conserved vectors
$\mathbf{\Lambda}_{(1)} = 1$ $\mathbf{T}^1_1 = w_t\ \mathbf{T}^2_1 = 3ww_x + w_{xxx} - w_x$	$\mathbf{\Lambda}^1_{(1)} = 1,\ \mathbf{\Lambda}^2_{(1)} = 0$ $R^1_1 = u_t$ $R^2_1 = -3vv_x + 3uu_x + u_{xxx} - u_x$ $I^1_1 = v_t$ $I^2_1 = 3uv_x + 3u_xv + v_{xxx} - v_x$
$\mathbf{\Lambda}_{(2)} = t$ $\mathbf{T}^1_2 = -w + tw_t$ $\mathbf{T}^2_2 = 3tww_x + tw_{xxx} - tw_x$	$\mathbf{\Lambda}^1_{(2)} = t,$ $\mathbf{\Lambda}^2_{(2)} = 0$ $R^1_2 = -u + tu_t$ $R^2_2 = t(3uu_x - 3vv_x + u_{xxx} - u_x)$ $I^1_2 = -v + tv_t$ $I^2_2 = t(3uv_x + 3vu_x + v_{xxx} - v_x)$
$\mathbf{\Lambda}_{(3)} = x$ $\mathbf{T}^1_3 = xw_t$ $\mathbf{T}^2_3 = 3xww_x - \frac{3}{2}w^2$ $+w + xw_{xxx} - xw_x - w_{xx}$	$\mathbf{\Lambda}^1_{(3)} = x,\ \mathbf{\Lambda}^2_{(3)} = 0$ $R^1_3 = xu_t$ $R^2_3 = -3xvv_x - \frac{3}{2}u^2 + 3xuu_x + \frac{3}{2}v^2 + u$ $+xu_{xxx} - xu_x - u_{xx}$ $I^1_3 = xv_t$ $I^2_3 = 3xuv_x + 3xvu_x - 3uv + xv_{xxx}$ $+v - xv_x - v_{xx}$
$\mathbf{\Lambda}_{(4)} = xt$ $\mathbf{T}^1_4 = -xw + xtw_t$ $\mathbf{T}^2_4 = 3xtww_x - \frac{3}{2}tw^2$ $+tw + xtw_{xxx} - xtw_x - tw_{xx}$	$\mathbf{\Lambda}^1_{(4)} = xt,\ \mathbf{\Lambda}^2_{(4)} = 0$ $R^1_4 = -xu + xtu_t$ $R^2_4 = -3xtvv_x - \frac{3}{2}tu^2 + 3xtuu_x$ $+\frac{3}{2}tv^2 + tu + xtu_{xxx} - xtu_x - tu_{xx}$ $I^1_4 = -xv + xtv_t$ $I^2_4 = 3xtuv_x + 3xtvu_x - 3tuv$ $+xtv_{xxx} + tv - xtv_x - tv_{xx}$

(continued)

Table 2 (continued)

The Boussinesq equation: $w_{tt} - w_{xx} + 3ww_{xx} + 3w_x^2 + w_{xxxx} = 0$	
Complex multipliers and conserved vectors	Split multipliers and conserved vectors
$u_{tt} - u_{xx} + 3uu_{xx} - 3vv_{xx} + 3u_x^2 - 3v_x^2 + u_{xxxx} = 0$ $v_{tt} - v_{xx} + 3vu_{xx} + 3uv_{xx} + 6u_xv_x + v_{xxxx} = 0$	
Multipliers for the split system	Conserved vectors for the split system
$\Lambda^1_{(1)} = 1,\ \Lambda^2_{(1)} = 0$	$T_1^1 = u_t$ $T_1^2 = -3vv_x + 3uu_x + u_{xxx} - u_x$
$\Lambda^1_{(2)} = 0,\ \Lambda^2_{(2)} = 1$	$T_2^1 = v_t$ $T_2^2 = 3uv_x + 3u_xv + v_{xxx} - v_x$
$\Lambda^1_{(3)} = t$	$T_3^1 = -u + tu_t$
$\Lambda^2_{(3)} = 0$	$T_3^2 = t(3uu_x - 3vv_x + u_{xxx} - u_x)$
$\Lambda^1_{(4)} = 0,\ \Lambda^2_{(4)} = t$	$T_4^1 = -v + tv_t$ $T_4^2 = t(3uv_x + 3vu_x + v_{xxx} - v_x)$
$\Lambda^1_{(5)} = x,\ \Lambda^2_{(5)} = 0$	$T_5^1 = xu_t$ $T_5^2 = -3xvv_x - \frac{3}{2}u^2 + 3xuu_x + \frac{3}{2}v^2 + u$ $+xu_{xxx} - xu_x - u_{xx}$
$\Lambda^1_{(6)} = 0,\ \Lambda^2_{(6)} = x$	$T_6^1 = xv_t$ $T_6^2 = 3xuv_x + 3xvu_x - 3uv + xv_{xxx}$ $+v - xv_x - v_{xx}$
$\Lambda^1_{(7)} = xt,\ \Lambda^2_{(7)} = 0$	$T_7^1 = -xu + xtu_t$ $T_7^2 = -3xtvv_x - \frac{3}{2}tu^2 + 3xtuu_x$ $+\frac{3}{2}tv^2 + tu + xtu_{xxx} - xtu_x - tu_{xx}$
$\Lambda^1_{(8)} = 0,\ \Lambda^2_{(8)} = xt$	$T_8^1 = -xv + xtv_t$ $T_8^2 = 3xtuv_x + 3xtvu_x - 3tuv$ $+xtv_{xxx} + tv - xtv_x - tv_{xx}$

3.4 The Complex Wave Equation

We apply the multiplier method to the wave equation in the complex domain. From Table 3, we see that the decomposed multipliers are different from the multipliers of the decomposed system. In this split case we obtain more conserved vectors. The conserved vector (T_7^1, T_7^2) is not deduced from the split conserved vectors of the wave equation with dissipative terms in the complex domain. Thus if the decomposed system is uncoupled, the conserved vectors for the decomposed system and split conserved vectors are not generally one to one. The arbitrary functions $\alpha(t, x)$ and $\beta(t, x)$ satisfy the wave equation with dissipation.

Table 3 Multipliers and conserved vectors of the nonlinear complex and split wave equation with dissipation

The wave equation with dissipation: $w_{tt} + w_t = w_{xx}$	
Complex multipliers and conserved vectors	Split multipliers and conserved vectors
$\mathbf{\Lambda}_{(1)} = e^t w_x$ $\mathbf{T}_1^1 = -e^t w_x w_t \ \mathbf{T}_1^2 = \frac{1}{2} e^t (w_t^2 + w_x^2)$	$\mathbf{\Lambda}_{(1)}^1 = e^t u_x, \ \mathbf{\Lambda}_{(1)}^2 = e^t v_x$ $R_1^1 = e^t(-u_t u_x + v_t v_x)$ $R_1^2 = \frac{1}{2} e^t (u_t^2 + u_x^2 - v_t^2 - v_x^2)$ $I_1^1 = -e^t(u_x v_t + u_t v_x)$ $I_1^2 = e^t(u_t v_t + u_x v_x)$
$\mathbf{\Lambda}_{(2)} = \frac{1}{2} e^t (2w_t + w)$ $\mathbf{T}_2^1 = -\frac{1}{2} e^t (w_t^2 + w_x^2 + w w_t)$ $\mathbf{T}_2^2 = e^t (\frac{1}{2} w w_x + w_x w_t)$	$\mathbf{\Lambda}_{(2)}^1 = \frac{1}{2} e^t (2u_t + u),$ $\mathbf{\Lambda}_{(2)}^2 = \frac{1}{2} e^t (2v_t + v)$ $R_2^1 = -\frac{1}{2} e^t (u u_t + u_t^2 + u_x^2 - v v_t - v_t^2 - v_x^2)$ $R_2^2 = \frac{1}{2} e^t (u u_x + 2u_t u_x - v v_x - 2v_t v_x)$ $I_2^1 = -\frac{1}{2} e^t (v u_t + u v_t + 2u_t v_t + 2u_x v_x)$ $I_2^2 = \frac{1}{2} e^t (v u_x + 2u_x v_t + u v_x + 2u_t v_x)$
$\mathbf{\Lambda}_{(3)} = \frac{1}{2} e^t (2x w_t + x w + 2t w_x)$ $\mathbf{T}_3^1 = -\frac{1}{2} e^t (x w_t^2 + x w_x^2$ $+ x w w_t + 2t w_t w_x)$ $\mathbf{T}_3^2 = \frac{1}{2} e^t (t w_t^2 + t w_x^2$ $+ x w w_x + 2x w_x w_t - \frac{1}{2} w^2)$	$\mathbf{\Lambda}_{(3)}^1 = \frac{1}{2} e^t (2x u_t + x u + 2t u_x),$ $\mathbf{\Lambda}_{(3)}^2 = \frac{1}{2} e^t (2x v_t + x v + 2t v_x)$ $R_3^1 = -\frac{1}{2} e^t (x u u_t + x u_t^2 + 2t u_t u_x$ $+ x u_x^2 - x v v_t - x v_t^2 - 2t v_t v_x - x v_x^2)$ $R_3^2 = \frac{1}{2} e^t (t u_t^2 + x u u_x + 2x u_t u_x + t u_x^2$ $- t v_t^2 - x v v_x - 2x v_t v_x - t v_x^2 + \frac{1}{2} v^2 - \frac{1}{2} u^2)$ $I_3^1 = -\frac{1}{2} e^t (x v u_t + x u v_t$ $+ 2x u_t v_t + 2t u_x v_t + 2t u_t v_x + 2x u_x v_x)$ $I_3^2 = \frac{1}{2} e^t (x v u_x + x u v_x$ $+ 2t u_t v_t + 2x u_x v_t + 2x u_t v_x + 2t u_x v_x - u v)$
$\mathbf{\Lambda}_\alpha = \alpha e^t$ $\mathbf{T}_\alpha^1 = e^t (\alpha w_t - \alpha_t w)$ $\mathbf{T}_\alpha^2 = e^t (-\alpha w_x + \alpha_x w)$	$\mathbf{\Lambda}_\alpha^1 = \alpha e^t, \ \mathbf{\Lambda}_\alpha^2 = 0$ $R_\alpha^1 = e^t (\alpha u_t - \alpha_t u)$ $R_\alpha^2 = e^t (-\alpha u_x + + u \alpha_x)$ $I_\alpha^1 = e^t (\alpha v_t - v \alpha_t)$ $I_\alpha^2 = e^t (-\alpha v_x + v \alpha_x)$
$u_{tt} + u_t - u_{xx} = 0, \ v_{tt} + v_t - v_{xx} = 0$	
Multipliers for the split system	Conserved vectors for the split system
$\Lambda_{(1)}^1 = -u_x e^t, \ \Lambda_{(1)}^2 = v_x e^t$	$T_1^1 = e^t(-u_t u_x + v_t v_x)$ $T_1^2 = \frac{1}{2} e^t (u_t^2 + u_x^2 - v_t^2 - v_x^2)$
$\Lambda_{(2)}^1 = v_x e^t, \ \Lambda_{(2)}^2 = u_x e^t$	$T_2^1 = -e^t (u_x v_t + u_t v_x)$ $T_2^2 = e^t (u_t v_t + u_x v_x)$
$\Lambda_{(3)}^1 = -\frac{1}{2}(u + 2u_t) e^t$	$T_3^1 = -\frac{1}{2} e^t (u u_t + u_t^2 + u_x^2$ $- v v_t - v_t^2 - v_x^2)$
$\Lambda_{(3)}^2 = \frac{1}{2}(v + 2v_t) e^t$	$T_3^2 = \frac{1}{2} e^t (u u_x + 2u_t u_x - v v_x - 2v_t v_x)$
$\Lambda_{(4)}^1 = \frac{1}{2}(v + 2v_t) e^t, \ \Lambda_{(4)}^2 = \frac{1}{2}(u + 2u_t) e^t$	$T_4^1 = -\frac{1}{2} e^t (v u_t + u v_t + 2u_t v_t + 2u_x v_x)$ $T_4^2 = \frac{1}{2} e^t (v u_x + 2u_x v_t + u v_x + 2u_t v_x)$
$\Lambda_{(5)}^1 = -\frac{1}{2}(x u + 2x u_t + 2t u_x) e^t$	$T_5^1 = -\frac{1}{2} e^t (x u u_t + x u_t^2 + 2t u_t u_x$ $+ x u_x^2 - x v v_t - x v_t^2 - 2t v_t v_x - x v_x^2)$

(continued)

Table 3 (continued)

The wave equation with dissipation: $w_{tt} + w_t = w_{xx}$	
Complex multipliers and conserved vectors	Split multipliers and conserved vectors
$\Lambda^2_{(5)} = \frac{1}{2}(xv + 2xv_t + 2tv_x)e^t$	$T^2_5 = \frac{1}{2}e^t(tu_t^2 + xuu_x + 2xu_tu_x + tu_x^2$ $-tv_t^2 - xvv_x - 2xv_tv_x - tv_x^2 + \frac{1}{2}v^2 - \frac{1}{2}u^2)$
$\Lambda^1_{(6)} = \frac{1}{2}(xv + 2xv_t + 2tv_x)e^t$	$T^1_6 = -\frac{1}{2}e^t(xvu_t + xuv_t + 2xu_tv_t$ $+2tu_xv_t + 2tu_tv_x + 2xu_xv_x)$
$\Lambda^2_{(6)} = \frac{1}{2}(xu + 2xu_t + 2tu_x)e^t$	$T^2_6 = \frac{1}{2}e^t(xvu_x + xuv_x + 2tu_tv_t$ $+2xu_xv_t + 2xu_tv_x + 2tu_xv_x - uv)$
$\Lambda^1_{(7)} = ve^t,\ \Lambda^2_{(7)} = -ue^t$	$T^1_7 = 2e^t(-uv_t + vu_t)$ $T^2_7 = 2e^t(uv_x - vu_x)$
$\Lambda^1_\alpha = \alpha e^t,\ \Lambda^2_\alpha = 0$	$T^1_\alpha = e^t(\alpha_t u - \alpha u_t)$ $T^2_\alpha = e^t(-u\alpha_x + \alpha u_x)$
$\Lambda^1_\beta = 0,\ \Lambda^2_\beta = \beta e^t$	$T^1_\beta = e^t(\beta_t v - \beta v_t)$ $T^2_\beta = e^t(-v\beta_x + \beta v_x)$

4 Conclusion

A complex PDE was decomposed into a system of two real PDEs. The multiplier method was investigated for the derivation of conserved quantities for an rth-order complex PDE and their decomposed systems for both Lagrangian and non-Lagrangian systems. The multipliers of the decomposed system were derived by utilizing the multiplier method. The multipliers of the decomposed system were not in general identical to the multipliers of the decomposed system of real PDEs. We have shown the result when they are the same. We finally concluded that the decomposed conserved vectors of the complex PDE were identical in number to the conserved vectors of the decomposed system of real PDEs in the case of the coupled system for both Lagrangian and non-Lagrangian systems. For an uncoupled decomposed system all the decomposed conserved vectors were not in general one to one with the conserved vectors for the decomposed system of real PDEs.

Acknowledgements F.M. is grateful to the N.R.F. of South Africa for a research grant which has facilitated this work.

References

1. Naz, R., Mahomed, F.M., Mason, D.P.: Comparison of different approaches to conservation laws for some partial differential equations in fluid mechanics. Appl. Math. Comput. **205**(1), 212–230 (2008)
2. Anco, S.C., Bluman, G.: Direct construction of conservation laws from field equations. Phys. Rev. Lett. **78**(15), 2869–2873 (1997)

3. Anco, S.C., Bluman, G.: Direct construction method for conservation laws of partial differential equations part II: general treatment. Eur. J. Appl. Math. **13**(05), 567–585 (2002)
4. Ali, S., Mahomed, F.M., Qadir, A.: Complex Lie symmetries for variational problems. J. Nonlinear Math. Phys. **15**(sup1), 25–35 (2008)
5. Farooq, M.U., Ali, S., Mahomed, F.M.: Two-dimensional systems that arise from the Noether classification of Lagrangians on the line. Appl. Math. Comput. **217**(16), 6959–6973 (2011)
6. Ali, S., Mahomed, F.M., Qadir, A.: Complex Lie symmetries for scalar second-order ordinary differential equations. Nonlinear Anal.: Real. World Appl. **10**(6), 3335–3344 (2009)
7. Soh, C.W., Mahomed, F.M.: Hypercomplex analysis and integration of systems of ordinary differential equations. Math. Methods Appl. Sci. (2016)
8. Mahomed, F.M., Naz, R.: A note on the Lie symmetries of complex partial differential equations and their split real systems. Pramana **77**(3), 483–491 (2011)
9. Naz, R., Mahomed, F.M.: A complex Noether approach for variational partial differential equations. Commun. Nonlinear Sci. Numer. Simul. **27**(1), 120–135 (2015)
10. Cheviakov, A.F.: GeM software package for computation of symmetries and conservation laws of differential equations. Comput. Phys. Commun. **176**(1), 48–61 (2007)
11. Cheviakov, A.F.: Computation of fluxes of conservation laws. J. Eng. Math. **66**(1–3), 153–173 (2010)

The Calculation and Use of Generalized Symmetries for Second-Order Ordinary Differential Equations

C. Muriel, J. L. Romero and A. Ruiz

Abstract New relationships between generalized symmetries, commuting symmetries, and generalized $\mathscr{C}^\infty-$symmetries for second-order ordinary differential equations are established. The sets of solutions of the respective determining equations are interrelated, which provides new strategies for solving them. Particular solutions of these determining equations can be appropriately combined in order to provide first integrals and Jacobi last multipliers for the equation.

Keywords Differential equation · Generalized symmetry · First integral · Jacobi last multiplier

1 Introduction

In many of the cases where an exact solution of a differential equation can be found, some type of symmetry for the equation is involved. Symmetries are usually defined through some transformations of variables that leave a given differential equation invariant. For instance, Lie point symmetries of an ordinary differential equation (ODE) are related to point transformations locally defined in the space of the independent and dependent variables of the ODE which map solutions into solutions [1, 4, 13–15]. If the transformations are allowed to act not only over the independent and dependent variables but also over their derivatives, new classes of symmetries appear. This is the case, for instance, of contact symmetries and of generalized (or dynamical) symmetries. The more general the admissible transformations are, the

C. Muriel (✉) · J. L. Romero (✉) · A. Ruiz (✉)
Departmento de Matemáticas, Universidad de Cádiz, 11510 Puerto Real, Spain
e-mail: concepcion.muriel@uca.es

J. L. Romero
e-mail: juanluis.romero@uca.es

A. Ruiz
e-mail: adrian.ruiz@uca.es

V. G. Kac et al. (eds.), *Symmetries, Differential Equations and Applications*, Springer Proceedings in Mathematics & Statistics 266,
https://doi.org/10.1007/978-3-030-01376-9_8

more symmetries will exist, but the more difficult it will be to find and use them to integrate the equation [15].

It is well-known that Lie point symmetries always exist for first-order ODEs, although there is no systematic way to find all of them. Something similar happens to contact transformations for second-order ODEs: contact symmetries always exist but, in general, they cannot be found explicitly. In order to find particular solutions of the corresponding determining equations it is usual to assume some *ad hoc ansatz* and check if it leads to some solution. Whereas Lie point symmetries can be used to reduce the order of a given ODE in a systematic way, a similar procedure does not apply for contact or generalized symmetries. In fact, applications of generalized symmetries are very rare in the literature and most of the examples are concerned only with systems of ODEs derivable from a Lagrangian.

This paper is devoted to establish some relationships between several mathematical objects related to a given second-order ODE such as generalized symmetries, commuting symmetries, first integrals, $\mathscr{C}^\infty-$symmetries, Jacobi last multipliers etc. As a consequence, by using some of these objects, several integration procedures can be combined.

In Sect. 2, after fixing the notations and collecting some basic notions and preliminary results, it is considered the concept of equivalence between generalized symmetries that lets to restrict our study to a special subclass of generalized symmetries. This is initially justified in Sect. 3 through Theorem 2: a first relation between commuting symmetries and generalized $\mathscr{C}^\infty-$symmetries [7, 11] is stated, and it is used later to derive important properties of the structure of the sets of solutions of the determining equations for both commuting and generalized $\mathscr{C}^\infty-$symmetries.

In sect. 4 we investigate how to find some solutions to any of the determining equations for commuting and generalized $\mathscr{C}^\infty-$symmetries by using solutions of several equations related to the differential operator associated to the given ODE. In particular, it is shown how some solutions to these equations can be combined in various ways to provide first integrals and Jacobi last multipliers of the equation.

The theoretical results have been supplemented throughout the paper by several examples that, for the sake of clarity, have been chosen to show the many different integration strategies that can be followed combining some of the mentioned objects. Although some of the presented equations could also have been solved by classical methods, the procedures used in this paper are completely different and provide new strategies of integration, which can be used even for equations lacking Lie point symmetries.

2 Generalized Symmetries and Generalized $\mathscr{C}^\infty$–Symmetries

2.1 Notations and Some Concepts on Symmetries

Throughout this paper M will denote an open subset of the space of the independent and dependent variables (x, u) of the second-order ODE

$$u_2 = \phi(x, u, u_1), \tag{1}$$

where u_i denotes the derivative of order i of u with respect to x, $i = 1, 2$. Let

$$\Delta = \{(x, u, u_1, u_2) \in M^{(2)} : u_2 = \phi(x, u, u_1)\} \tag{2}$$

denote the manifold on $M^{(2)}$ defined by (1). The restriction of the total derivative operator $\mathbf{D}_x$ to Δ determines the vector field $\mathbf{A} = \partial_x + u_1\partial_u + \phi(x, u, u_1)\partial_{u_1}$ associated to (1).

In what follows in this section, we provide some basic definitions and results for three different concepts of symmetry for an equation (1). Some relationships between these concepts will be studied in Sect. 3.

A.- It is well-known [1, 13–15] that the ordinary differential equation (1) admits a Lie point symmetry with generator $\mathbf{v} = \xi(x, u)\partial_x + \eta(x, u)\partial_u$ if and only if

$$\mathbf{v}^{(2)}\big(u_2 - \phi(x, u, u_1)\big) = 0 \quad \text{mod.} \quad u_2 = \phi(x, u, u_1), \tag{3}$$

where $\mathbf{v}^{(2)}$ denotes the standard 2nd-order prolongation of $\mathbf{v}$. Equivalently, $\mathbf{v}$ is a Lie point symmetry of (1) if and only if

$$\left[\mathbf{v}^{(1)}, \mathbf{A}\right] = -\mathbf{A}(\xi)\mathbf{A} \tag{4}$$

holds [15].

B.- The generalized symmetries of (1) are generalized vector fields

$$\mathbf{v} = \xi(x, u, u_1)\partial_x + \eta(x, u, u_1)\partial_u \tag{5}$$

such that (3) holds. These symmetries can also be characterized as the vector fields (5) such that

$$\left[\mathbf{v}_\Delta^{(1)}, \mathbf{A}\right] = -\mathbf{A}(\xi) \cdot \mathbf{A}, \tag{6}$$

where $\mathbf{v}_\Delta^{(1)}$ denotes the restriction of $\mathbf{v}^{(1)}$ to Δ.

C.- Let $\mathbf{v} = \xi(x, u, u_1)\partial_x + \eta(x, u, u_1)\partial_u$ be a generalized vector field and let $\lambda_0 = \lambda_0(x, u, u_1)$ be a smooth function. The pair $(\mathbf{v}, \lambda_0)$ is called a *generalized* $\mathscr{C}^\infty(M^{(1)})$*–symmetry* (or briefly, a $\mathscr{C}^\infty$–symmetry) of equation (1) if the first-order

λ_0−prolongation of $\mathbf{v}$ [7, 11]

$$\mathbf{v}^{[\lambda_0,(1)]} = \mathbf{v} + \big((\mathbf{A}+\lambda_0)(\eta) - (\mathbf{A}+\lambda_0)(\xi)u_1\big)\partial u_1 \tag{7}$$

verifies

$$\left[\mathbf{v}^{[\lambda_0,(1)]}, \mathbf{A}\right] = \lambda_0 \cdot \mathbf{v}^{[\lambda_0,(1)]} - (\mathbf{A}+\lambda_0)(\xi)\cdot \mathbf{A}. \tag{8}$$

This is equivalent to the following invariance condition:

$$\mathbf{v}^{[\lambda_0,(2)]}(u_2 - \phi(x,u,u_1)) = 0 \quad \text{mod.} \quad u_2 = \phi(x,u,u_1). \tag{9}$$

The original concept of (standard) $\mathscr{C}^\infty$−symmetry [7] refers to the case when the infinitesimals ξ and η of $\mathbf{v}$ do not depend on u_1. In this case, the operator $\mathbf{A}$ can be replaced by $\mathbf{D}_x$ in (7). It is also clear from (4), (6) and (8) that Lie point (resp. generalized) symmetries correspond to standard (resp. generalized) $\mathscr{C}^\infty$−symmetries such that $\lambda_0 = 0$.

2.2 A−Equivalence

It must be observed that if (5) is a generalized symmetry of the equation (1) then, for an arbitrary smooth function $\zeta = \zeta(x,u,u_1)$, the generator $\mathbf{v}_\Delta^{(1)} + \zeta(x,u,u_1)\mathbf{A}$ is also related to a generalizaed symmetry of the equation (1); in particular, $\mathbf{v}_\Delta^{(1)} - \xi(x,u,u_1)\mathbf{A}$ corresponds to a generalized symmetry of the equation such that the coefficient of ∂_x is null. To get rid of this degree of freedom, we consider the following equivalence relationship:

Definition 1 We will say that two generalized vector fields $\mathbf{V}_1, \mathbf{V}_2$ on $M^{(1)}$ are $\mathbf{A}$−equivalent, and we will use the notation $\mathbf{V}_1 \overset{\mathbf{A}}{\sim} \mathbf{V}_2$, if and only if the vector fields $\mathbf{A}, \mathbf{V}_1, \mathbf{V}_2$ are linearly dependent over $\mathscr{C}^\infty(M^{(1)})$.

The concept of $\mathbf{A}$−equivalent vector fields can be translated to $\mathscr{C}^\infty$−symmetries by considering their first λ−prolongations: two $\mathscr{C}^\infty$−symmetries $(\mathbf{v}_1, \lambda_1)$ and $(\mathbf{v}_2, \lambda_2)$ of (1) are called $\mathbf{A}$−equivalent if $\mathbf{v}_1^{[\lambda_1,(1)]}$ and $\mathbf{v}_2^{[\lambda_2,(1)]}$ are $\mathbf{A}$−equivalent according to Definition 1. We refer to [11] for a detailed description of the significance of this notion.

If $\mathbf{v} = \xi(x,u,u_1)\partial_u + \eta(x,u,u_1)\partial_u$ and $(\mathbf{v},\lambda_0)$ is a $\mathscr{C}^\infty$−symmetry of (1), it can be checked [11] that $\mathbf{v}^{[\lambda_0,(1)]}$ has the form

$$\mathbf{v}^{[\lambda_0,(1)]} = Q(\partial_u)^{[\lambda,(1)]} + \xi\mathbf{A}, \tag{10}$$

where $Q = \eta - \xi u_1$ is the characteristic of $\mathbf{v}$ and

$$\lambda = \lambda_0 + \frac{\mathbf{A}(Q)}{Q}. \tag{11}$$

It can be proved [11] that this function λ satisfies the equation

$$\mathbf{A}(\lambda) + \lambda^2 = \phi_u + \lambda\phi_{u_1} \tag{12}$$

and conversely, if λ satisfies (12) then (∂_u, λ) is a $\mathcal{C}^\infty-$ symmetry of (1). As a consequence of (10), the vector fields $\mathbf{v}^{[\lambda_0,(1)]}$ and $(\partial_u)^{[\lambda,(1)]}$ are $\mathbf{A}-$equivalent. The pair (∂_u, λ) is also a $\mathcal{C}^\infty-$symmetry of (1) which will be called the *canonical representative* of the $\mathbf{A}-$equivalence class corresponding to $(\mathbf{v}, \lambda_0)$; this canonical representative is uniquely defined by (11).

3 Relationships Between Generalized Symmetries and Generalized $\mathcal{C}^\infty-$Symmetries

In this section we derive several characterizations and relationships between the concepts considered in Sect. 2.

3.1 Commuting Symmetries

Let us recall that the evolutionary form of a given generalized symmetry $\mathbf{v} = \xi(x, u, u_1)\partial_x + \eta(x, u, u_1)\partial_u$ of (1) is the vector field $\mathbf{v}_Q = Q\partial_u$, where $Q = \eta - \xi u_1$ is the characteristic of $\mathbf{v}$. The following properties of $\mathbf{v}_Q$ can be easily proven:

1. $\mathbf{v}_Q(x) = 0$,
2. $[\mathbf{v}_Q^{(1)}, \mathbf{A}] = 0$,
3. $\mathbf{v}_Q^{(1)} = Q\partial_u + \mathbf{A}(Q)\partial_{u_1}$,
4. $\mathbf{A}^2(Q) = \phi_{u_1}\mathbf{A}(Q) + \phi_u Q$, where $\mathbf{A}^2(Q) = \mathbf{A}\big(\mathbf{A}(Q)\big)$.

For the reasons considered at the beginning of Sect. 2.2, in most studies on generalized symmetries the attention is focused on symmetries whose generators have no ∂_x term. Let us observe that if $\mathbf{v} = \xi(x, u, u_1)\partial_x + \eta(x, u, u_1)\partial_u$ is a generalized vector field, the condition $\mathbf{v}(x) = 0$ is equivalent to the condition $\xi = 0$. In this case, $\mathbf{v}$ is a generalized symmetry of (1) if and only if $\left[\mathbf{v}_\Delta^{(1)}, \mathbf{A}\right] = 0$. In this paper the term *commuting symmetry* will refer to vector fields $\mathbf{V}$ on $M^{(1)}$ of the form $\mathbf{V} = \xi(x, u, u_1)\partial_x + \eta(x, u, u_1)\partial_u + \eta_1(x, u, u_1)\partial_{u_1}$ that have properties similar to those satisfied by $\mathbf{v}_Q^{(1)}$. More precisely:

Definition 2 We will say that a vector field $\mathbf{V}$ defined on $M^{(1)}$ is a *commuting symmetry* of equation (1) if $\mathbf{V}(x) = 0$ and $[\mathbf{V}, \mathbf{A}] = 0$.

For further references, we include the following proposition, which can be easily proved by using the properties of the Lie bracket; it provides a characterization of commuting symmetries in terms of a function $f = f(x, u, u_1)$.

Proposition 1 *A vector field* $\mathbf{V}$ *on* $M^{(1)}$ *is a commuting symmetry of equation (1) if and only if* $\mathbf{V}$ *can be written as*

$$\mathbf{V} = f\partial_u + \mathbf{A}(f)\partial_{u_1} \tag{13}$$

where $f = f(x, u, u_1)$ *is a solution of*

$$\mathbf{A}^2(f) = \phi_{u_1}\mathbf{A}(f) + \phi_u f. \tag{14}$$

In what follows Eq. (14) will be called determining equation for commuting symmetries.

Although the set of generalized symmetries of an equation (1) strictly contains the set of commuting symmetries, next we prove that the first prolongation of any generalized symmetry of (1) is $\mathbf{A}-$equivalent to a commuting symmetry:

Proposition 2 *If* $\mathbf{v}$ *is a generalized symmetry of equation (1) then there exists a commuting symmetry* $\mathbf{V}$ *which is* $\mathbf{A}-$*equivalent to* $\mathbf{v}_{\Delta}^{(1)}$. *Conversely, if* $\mathbf{V}$ *is a commuting symmetry of (1) then there exist a generalized symmetry* $\mathbf{v}$ *such that* $\mathbf{V} = \mathbf{v}_{\Delta}^{(1)}$.

Proof Let $\mathbf{v} = \xi(x, u, u_1)\partial_x + \eta(x, u, u_1)\partial_u$ be a generalized symmetry of the equation (1) and let $\mathbf{v}_Q^{(1)}$ denote the restriction to Δ of the first prolongation of the evolutionary form of $\mathbf{v}$. It can be checked that

$$\mathbf{v}_{\Delta}^{(1)} = \mathbf{v}_Q^{(1)} + \xi\mathbf{A}$$

holds, which implies that $\mathbf{v}_{\Delta}^{(1)}$ and $\mathbf{v}_Q^{(1)}$ are $\mathbf{A}-$equivalent. Since $[\mathbf{v}_Q^{(1)}, \mathbf{A}] = 0$, the vector field $\mathbf{V} = \mathbf{v}_Q^{(1)}$ is a commuting symmetry of equation (1).

Conversely, let $\mathbf{V}$ be a commuting symmetry of equation (1). By Proposition 1, we can write $\mathbf{V} = f\partial_u + \mathbf{A}(f)\partial_{u_1}$, where $f = f(x, u, u_1)$ satisfies the equation (14). We define the (generalized) vector field $\mathbf{v} = f\partial_u$. Clearly $\mathbf{v}_{\Delta}^{(1)} = \mathbf{V}$, which implies that $[\mathbf{v}_{\Delta}^{(1)}, \mathbf{A}] = 0$, $\big($i.e. $\mathbf{v}$ is a generalized symmetry of (1)$\big)$.

3.2 Relationships Between Commuting Symmetries and $\mathscr{C}^\infty-$Symmetries

Let us suppose that $\mathbf{V} = f\partial_u + \mathbf{A}(f)\partial_{u_1}$ is a commuting symmetry of (1), where $f = f(x, u, u_1)$ is a non-zero solution for (14). Let $\lambda = \lambda(x, u, u_1)$ be the function defined by

$$\lambda = \frac{\mathbf{A}(f)}{f}. \tag{15}$$

Then $\mathbf{V}$ can be written as $\mathbf{V} = f(\partial_u + \lambda\partial_{u_1})$ and, by using (14), we can write

$$\mathbf{A}(\lambda) = \frac{\mathbf{A}\big(\mathbf{A}(f)\big)f - \mathbf{A}(f)\mathbf{A}(f)}{f^2} = \frac{\big(\phi_u f + \phi_{u_1}\mathbf{A}(f)\big)f - (\lambda f)(\lambda f)}{f^2}$$
$$= \phi_u + \phi_{u_1}\lambda - \lambda^2.$$

This proves that the function λ given by (15) satisfies

$$\mathbf{A}(\lambda) = \phi_u + \phi_{u_1}\lambda - \lambda^2, \tag{16}$$

which coincides with Eq. (12) and therefore (∂_u, λ) is a $\mathcal{C}^\infty$−symmetry of (1) [11]. Throughout this paper Eq. (16) will be called determining equation for $\mathcal{C}^\infty$−symmetries.

Conversely, let us assume that λ is an arbitrary solution of (16) and let $f = f(x, u, u_1)$ be any function such that

$$\mathbf{A}(f) = \lambda f. \tag{17}$$

Then, by using (16), we can write

$$\mathbf{A}^2(f) = \mathbf{A}(\mathbf{A}(f)) = \mathbf{A}(\lambda f) = \mathbf{A}(\lambda)f + \lambda\mathbf{A}(f) = \left(\phi_u + \phi_{u_1}\lambda - \lambda^2\right)f + \lambda(\lambda f)$$
$$= \phi_u f + \phi_{u_1}(\lambda f) = \phi_u f + \phi_{u_1}\mathbf{A}(f).$$

Therefore, f satisfies the second-order linear partial differential equation (14) and the vector field $\mathbf{V} = f(\partial_u + \lambda\partial_{u_1})$ is a commuting symmetry of (1).

We have proved the result that follows, which shows the relationships between commuting symmetries and generalized $\mathcal{C}^\infty$−symmetries.

Theorem 1 *If (∂_u, λ) is a $\mathcal{C}^\infty$−symmetry of the equation (1) then for any function f such that $\mathbf{A}(f) = \lambda f$ the vector field*

$$\mathbf{V} = f(\partial_u + \lambda\partial_{u_1}) = f(\partial_u)^{[\lambda,(1)]}$$

is a commuting symmetry of (1). Equivalently, if λ satisfies (16) and f is such that $\mathbf{A}(f) = \lambda f$ then f satisfies (14).

Conversely, if $\mathbf{V} = f\partial_u + \mathbf{A}(f)\partial_{u_1}$ is a commuting symmetry of (1) then the function λ defined by (15) is such that (∂_u, λ) is a $\mathcal{C}^\infty$−symmetry of (1). Equivalently, if f satisfies (14) then the function λ defined by (15) satisfies (16).

Remark 1 Commuting symmetries of (1) are determined by the solutions of (14), which is a 2nd-order linear partial differential equation. The solutions for this equation can be obtained from the solutions λ and f of the first-order partial differential equations (16) and (17), respectively. Whereas $\mathbf{A}(f) = \lambda f$ is a linear equation, the determining equation (16) for the $\mathscr{C}^\infty$−symmetries of (1) is a quasilinear first-order PDE. Therefore the commuting symmetries of (1) could be found by obtaining a solution λ of (16), a non null solution f for the equation $\mathbf{A}(f) = \lambda f$ and by using Theorem 1.

3.3 Form of the Solutions of the Determining Equation (14) for Commuting Symmetries

It this section we analyse the form of the solutions of the determining equation (14) for commuting symmetries.

Let f_1 and f_2 be two solutions of (14) and let $\mathbf{V}_1, \mathbf{V}_2$ be the corresponding commuting symmetries. Let us suppose that the set of vector fields $\{\mathbf{A}, \mathbf{V}_1, \mathbf{V}_2\}$ is linearly independent. This means that

$$W(f_1, f_2) = \begin{vmatrix} f_1 & \mathbf{A}(f_1) \\ f_2 & \mathbf{A}(f_2) \end{vmatrix} = \begin{vmatrix} 1 & u_1 & \phi \\ 0 & f_1 & \mathbf{A}(f_1) \\ 0 & f_2 & \mathbf{A}(f_2) \end{vmatrix} \neq 0, \tag{18}$$

which implies that $\mathbf{V}_1$ is not $\mathbf{A}$−equivalent to $\mathbf{V}_2$. Let I_1, I_2 be two arbitrary first integrals of $\mathbf{A}$. The function f defined by $f = I_1 f_1 + I_2 f_2$ satisfies

$$\begin{aligned} \mathbf{A}(f) &= \mathbf{A}(I_1) f_1 + I_1 \mathbf{A}(f_1) + \mathbf{A}(I_2) f_2 + I_2 \mathbf{A}(f_2) = I_1 \mathbf{A}(f_1) + I_2 \mathbf{A}(f_2), \\ \mathbf{A}^2(f) &= \mathbf{A}(I_1)\mathbf{A}(f_1) + I_1 \mathbf{A}^2(f_1) + \mathbf{A}(I_2)\mathbf{A}(f_2) + I_2 \mathbf{A}^2(f_2) \\ &= I_1 \mathbf{A}^2(f_1) + I_2 \mathbf{A}^2(f_2) \\ &= I_1(\phi_u f_1 + \phi_{u_1} \mathbf{A}(f_1)) + I_2(\phi_u f_2 + \phi_{u_1} \mathbf{A}(f_2)) \\ &= \phi_u (I_1 f_1 + I_2 f_2) + \phi_{u_1} \mathbf{A}(I_1 f_1 + I_2 f_2) = \phi_u f + \phi_{u_1} \mathbf{A}(f). \end{aligned} \tag{19}$$

Therefore, $f = I_1 f_1 + I_2 f_2$ is also a solution of (14).

We next prove that if f_1 and f_2 are two solutions of (14) such that (18) holds then any solution f to (14) can be written in the form $f = I_1 f_1 + I_2 f_2$ for some first integrals I_1, I_2 of $\mathbf{A}$. In order to motivate our proof, let us observe that in such case we would also have $\mathbf{A}(f) = I_1 \mathbf{A}(f_1) + I_2 \mathbf{A}(f_2)$ and $\lambda f = \lambda_1 I_1 f_1 + \lambda_2 I_2 f_2$, where $\lambda_1 = \mathbf{A}(f_1)/f_1$, $\lambda_2 = \mathbf{A}(f_2)/f_2$ and $\lambda = \mathbf{A}(f)/f$. Hence, the pair I_1, I_2 would be a solution of the linear system

$$\begin{aligned} f_1 I_1 + \quad f_2 I_2 &= \ f, \\ \lambda_1 f_1 I_1 + \lambda_2 f_2 I_2 &= \lambda f. \end{aligned} \tag{20}$$

Let us observe that, by (18), the determinant of the system (20) is $W(f_1, f_2) \neq 0$ and, by Cramer's rule, I_1, I_2 would be uniquely given by

$$I_1 = \frac{f(\lambda_2 - \lambda)}{f_1(\lambda_2 - \lambda_1)}, \quad I_2 = \frac{f(\lambda - \lambda_1)}{f_2(\lambda_2 - \lambda_1)}. \tag{21}$$

After this motivation that provides us with the form of the first integrals I_1, I_2 of $\mathbf{A}$, let us proceed to the proof of the announced result.

If f_1 and f_2 are two functions such that (18) holds then, for any solution f to (14), the functions I_1, I_2 in (21) are well defined and they satisfy $f = I_1 f_1 + I_2 f_2$. Since λ, λ_1 and λ_2 are solutions of (16), it can directly be checked that $\mathbf{A}(I_1) = \mathbf{A}(I_2) = 0$; i.e., I_1, I_2 are first integrals of $\mathbf{A}$.

As a consequence, the following theorem shows the forms of commuting symmetries and the solutions of (14):

Theorem 2 *Let f_1, f_2 be two solutions of (14) such that (18) holds and let $\mathbf{V}_i = f_i \partial_u + \mathbf{A}(f_i)\partial_{u_1}$, $i = 1, 2$, be the corresponding commuting symmetries. Then:*

1. *If f is an arbitrary solution of (14) and $\mathbf{V} = f\partial_u + \mathbf{A}(f)\partial_{u_1}$ is the corresponding commuting symmetry then there exist two first integrals I_1, I_2 of $\mathbf{A}$ such that $f = I_1 f_1 + I_2 f_2$ and $\mathbf{V} = I_1\mathbf{V}_1 + I_2\mathbf{V}_2$.*
2. *Conversely, if a function $f = f(x, u, u_1)$ can be written in the form $f = I_1 f_1 + I_2 f_2$, where I_1, I_2 are first integrals of $\mathbf{A}$, then f solves (14).*

3.4 Form of the Solutions of the Determining Equation (16) for $\mathscr{C}^\infty$–Symmetries

Theorems 1 and 2 can be used to determine the form of the solutions of (16). Let us suppose that λ, λ_1 and λ_2 are three solutions of (16) such that $\lambda_1 \neq \lambda_2$. Let f, f_1, f_2 be such that $\mathbf{A}(f) = \lambda f$, $\mathbf{A}(f_1) = \lambda_1 f_1$ and $\mathbf{A}(f_2) = \lambda_2 f_2$. By Theorem 2, there exist two first integrals I_1, I_2 of $\mathbf{A}$ such that (20) holds. Therefore

$$\begin{aligned} \lambda = \frac{\mathbf{A}(f)}{f} &= \lambda_1 \frac{I_1 f_1}{I_1 f_1 + I_2 f_2} + \lambda_2 \frac{I_2 f_2}{I_1 f_1 + I_2 f_2} \\ &= \lambda_1 \frac{1}{1 + (I_2 f_2)/(I_1 f_1)} + \lambda_2 \frac{(I_2 f_2)/(I_1 f_1)}{1 + (I_2 f_2)/(I_1 f_1)} = \frac{1}{1+h}\lambda_1 + \frac{h}{1+h}\lambda_2, \end{aligned} \tag{22}$$

where h is defined by $h = \dfrac{I_2 f_2}{I_1 f_1}$ and satisfies

$$\mathbf{A}(h) = (\lambda_2 - \lambda_1)h. \tag{23}$$

As a consequence, the following theorem holds.

Theorem 3 *Let (∂_u, λ), (∂_u, λ_1) and (∂_u, λ_2) be three $\mathscr{C}^\infty$–symmetries of (1). If $\lambda_1 \neq \lambda_2$ then the function h defined by $h = \dfrac{\lambda_1 - \lambda}{\lambda - \lambda_2}$ satisfies*

$$\lambda = \frac{1}{1+h}\lambda_1 + \frac{h}{1+h}\lambda_2 \tag{24}$$

and the equation $\mathbf{A}(h) = (\lambda_2 - \lambda_1)h$.

Conversely, if (∂_u, λ_1) *and* (∂_u, λ_2) *are two different* $\mathscr{C}^\infty$*−symmetries of (1) and* h *is any solution to the equation* $\mathbf{A}(h) = (\lambda_2 - \lambda_1)h$ *then the function* λ *given by (24) is such that* (∂_u, λ) *is a* $\mathscr{C}^\infty$*−symmetry of (1).*

3.5 An Application of Theorem 3 to the Equations in the Class $\mathscr{A}_1$

In this subsection we use Theorem 3 to determine all the $\mathscr{C}^\infty$−symmetries (∂_u, λ) admitted by a family of second-order equations that has been extensively studied in the recent literature. Such class of equations has been denoted by $\mathscr{A}_1$ in [8] and contains equations of the form

$$u_2 + a_2(x,u)u_1^2 + a_1(x,u)u_1 + a_0(x,u) = 0 \tag{25}$$

whose coefficients a_2, a_1, a_0 satisfy the following conditions [2, 8] :

$$\begin{aligned} S_1 &= a_{1u} - 2a_{2x} = 0, \\ S_2 &= (a_0a_2 + a_{0u})_u + (a_{2x} - a_{1u})_x + (a_{2x} - a_{1x})a_1 = 0. \end{aligned}$$

Relevant properties of the equations in this class have been deeply studied concerning, for instance, to their linearisation through both local and nonlocal transformations, as well as to the admitted first integrals and $\mathscr{C}^\infty$− symmetries [2, 8, 9].

In particular, the equations in the class $\mathscr{A}_1$ can be characterized as the Eq. (25) that admit two functionally independent first integrals of the form $I_i = A_i(x,u)u_1 + B_i(x,u)$, for $i = 1, 2$. The associated $\mathscr{C}^\infty$−symmetries (∂_u, λ_1) and (∂_u, λ_1) are defined by functions λ_1 and λ_2 that are also linear in u_1 (see Theorem 4 in [8] for details). The search of such first integrals and $\mathscr{C}^\infty$−symmetries is based on the construction of the function $f = f(x)$ defined by

$$f(x) = a_0a_2 + a_{0u} - \frac{1}{2}a_{1x} - \frac{1}{4}{a_1}^2, \tag{26}$$

which does not depend on u [8].

Let $g_1 = g_1(x)$, $g_2 = g_2(x)$ be a fundamental set of solutions of the linear second-order ODE $g''(x) + f(x)g(x) = 0$. According to [8], the functions

$$\begin{aligned} \lambda_1(x,u,u_1) &= -a_2(x,u)u_1 - \frac{a_1(x,u)}{2} + \frac{g_1'(x)}{g_1(x)}, \\ \lambda_2(x,u,u_1) &= -a_2(x,u)u_1 - \frac{a_1(x,u)}{2} + \frac{g_2'(x)}{g_2(x)}, \end{aligned} \tag{27}$$

are two different solutions of (16). By the other hand, it can be checked that $h_0 = g_2/g_1$ solves the equation $\mathbf{A}(h) = (\lambda_2 - \lambda_1)h$ and therefore any solution to this linear equation is of the form

$$h = I \cdot \frac{g_2}{g_1}, \tag{28}$$

where $I = I(x, u, u_1)$ is some first integral of $\mathbf{A}$.

The general solution of the corresponding determining equation (16) for $\mathcal{C}^\infty-$symmetries is given by (24). However, it is important to observe that it can be checked, by using the function h given in (28) and some simplifications, that the general solution of the corresponding equation (16) can also be written in the form

$$\lambda = -a_2(x,u)u_1 - \frac{a_1(x,u)}{2} + \frac{g_1'(x) + I g_2'(x)}{g_1(x) + I g_2(x)}, \tag{29}$$

where I is a first integral of $\mathbf{A}$.

Example 1 As an example of an equation in the class $\mathscr{A}_1$, we will consider the Morse equation

$$u_2 + u_1^2 + 1 - e^{-u} = 0. \tag{30}$$

For this equation the corresponding function (26) becomes $f(x) = 1$. Two independent solutions of the corresponding linear equation $g'' + fg = 0$ are $g_1(x) = \sin x$, $g_2(x) = \cos x$. According with the previous discussion, two different $\mathcal{C}^\infty-$symmetries (∂_u, λ_2) and (∂_u, λ_3) are given by the following particular solutions of (16)

$$\lambda_2 = -u_1 + \cot x \qquad \text{and} \qquad \lambda_3 = -u_1 - \tan x, \tag{31}$$

which satisfy $\lambda_3 - \lambda_2 = -\tan x - \cot x$. The solutions of $\mathbf{A}(h) = (\lambda_3 - \lambda_2)h$ are of the form $h = I \cdot \cot x$, where I is a first integral of (30). In consequence, any $\mathcal{C}^\infty-$symmetry (∂_u, λ) of (30) is determined by a function λ of the form

$$\lambda = -u_1 + \frac{\cos x - I \cdot \sin x}{\sin x + I \cdot \cos x}, \tag{32}$$

where I is a first integral of $\mathbf{A}$.

4 Combined Use of Solutions of the Determining Equations for Commuting Symmetries and $\mathscr{C}^\infty$–Symmetries

In this section we show how some solutions of any of the Eqs. (14) or (16) can be used to obtain solutions of several equations related to the differential operator **A**.

4.1 Reductions of the Determining Equations (16) and (14)

As it has been mentioned in Remark 1, the solutions f for (14) can be obtained by solving first the equation (16) and then by solving an equation of the form $\mathbf{A}(f) = \lambda f$, where λ is an arbitrary solution of (16). In this subsection we analyse the structure of the solutions of equation (16) by following a strategy similar to the classical transformation of a Riccati-type equation into a Bernoulli-type equation and a linear one [5].

Let us assume that $\lambda_1 = \lambda_1(x, u, u_1)$ is a known particular solution of (16). It can be checked that if λ is an arbitrary solution of (16) then the function $\mu = \lambda - \lambda_1$ is a solution of

$$\mathbf{A}(\mu) = (-2\lambda_1 + \phi_{u_1})\mu - \mu^2. \tag{33}$$

Conversely, it can also be checked that if μ is an arbitrary solution of (33) then the function $\lambda = \mu + \lambda_1$ is a solution of (16).

The equation (33) is a Bernoulli-type equation; if $\mu = \mu(x, u, u_1)$ is a non-zero solution of (33) then $\nu = 1/\mu$ satisfies the (non-homogeneous) linear equation

$$\mathbf{A}(\nu) = (2\lambda_1 - \phi_{u_1})\nu + 1. \tag{34}$$

Conversely, if $\nu = \nu(x, u, u_1)$ is a non-zero solution of (34) then $\mu = 1/\nu$ is a solution of (33).

In order to transform (34) into an homogeneous linear equation, let us assume that ν_1 is a particular solution of (34). It can be easily checked that a function ν is a solution of (34) if and only if the function $\zeta = \nu - \nu_1$ is a solution of the linear homogeneous equation

$$\mathbf{A}(\zeta) = (2\lambda_1 - \phi_{u_1})\zeta. \tag{35}$$

Let us assume that $\zeta_1 = \zeta_1(x, u, u_1)$ is a particular solution of (35). Since (35) is a linear equation, if I is a first integral of **A** then $\zeta(x, u, u_1) = I\zeta_1$ is a solution of (35) and conversely: any solution ζ of (35) is of the form $\zeta(x, u, u_1) = I\zeta_1$, where I is a first integral of **A**.

Previous discussion shows that any solution of (16) can be expressed in terms of a particular solution λ_1 of (16), a particular solution ν_1 of the linear (non-homogeneous) equation (34) and a solution of the linear equation (35):

Proposition 3 *Let λ_1, ν_1 and ζ_1 be particular solutions of the equations (16), (34) and (35), respectively. If $I = I(x, u, u_1)$ is any first integral of* **A** *then the function*

$$\lambda = \frac{1}{I\zeta_1 + \nu_1} + \lambda_1 \tag{36}$$

is a solution of (16). Conversely, if λ is any solution to (16) then the function I defined through (36) as

$$I = \frac{1}{\zeta_1}\left(\frac{1}{\lambda - \lambda_1} - \nu_1\right) \tag{37}$$

is a first integral of **A**.

Remark 2 There are several circumstances wherein particular solutions ν_1 or ζ_1 of (34) or (35), respectively, can be obtained by using several classes of solutions of equations associated to the differential operator **A** of the given equation.

As an example, let us observe that if λ_2 is a second known particular solution of (16) then $\mu_1 = \lambda_2 - \lambda_1$ is a particular solution of (33). If μ_1 is a non-zero function then $\nu_1 = 1/\mu_1 = 1/(\lambda_2 - \lambda_1)$ is a particular solution of (34). Therefore, if λ is a solution to (16) written in the form (36), the expression $\nu_1 = 1/(\lambda_2 - \lambda_1)$ is used in (36) and the resulting expression is simplified then we get that the function λ can also be written as

$$\lambda = \frac{1}{1+h}\lambda_1 + \frac{h}{1+h}\lambda_2, \tag{38}$$

where

$$h = \frac{1}{(\lambda_2 - \lambda_1)I\zeta_1} \tag{39}$$

satisfies $\mathbf{A}(h) = (\lambda_2 - \lambda_1)h$. These expressions should be compared with (24).

Conversely, if a function $\lambda = \lambda(x, u, u_1)$ can be written in the form (38)–(39), where λ_1 and λ_2 are particular solutions of (16), and h satisfies the equation $\mathbf{A}(h) = (\lambda_2 - \lambda_1)h$ then a direct calculation shows that λ is a solution of (16).

Therefore, Theorem 3 could also have been proved by using the discussion in this subsection.

Remark 3 As we have mentioned before, Theorem 3 is similar to a well-known result for a Riccati equation [5]

$$y' = f_2(x)y^2 + f_1(x)y + f_0(x). \tag{40}$$

If $y_1(x)$ and $y_2(x)$ are two solutions of this equation then the general solution of the Riccati equation can be written in a form similar to (38):

$$y = \frac{Cy_1 + U(x)y_2}{C + U(x)} = \frac{1}{1+h}y_1 + \frac{h}{1+h}y_2,$$

where $U(x) = \exp\left(\int f_2(x)\big(y_1(x) - y_2(x)\big)dx\right)$, C is an arbitrary constant and $h = U/C$.

Let also recall that another well-known property of a Riccati equation (40) is that it can be transformed into a second-order linear equation by means of the new dependent variable given by

$$u(x) = \exp\left(-\int f_2(x)y(x)dx\right).$$

At the beginning of this section we have shown that a similar result holds for the Eq. (16), by considering the differential operator $\mathbf{A}$ instead of the derivative y'.

Remark 4 Let us recall that a function $M = M(x, u, u_1)$ such that

$$\mathbf{A}(M) = -\phi_{u_1} M$$

is known in the literature with the name of Jacobi last multiplier for (1) [12]. In this remark we show how Jacobi last multipliers are narrowly linked to the solutions of (14), (16) and (35).

For instance, if M is a Jacobi last multiplier for (1), λ_1 is a particular solution of (16) and f satisfies $\mathbf{A}(f) = \lambda_1 f$, then $\mathbf{A}\left(f^2\right) = 2\lambda_1 f^2$ and $\zeta_1 = Mf^2$ satisfies

$$\begin{aligned}\mathbf{A}(\zeta_1) &= \mathbf{A}\left(Mf^2\right) = \mathbf{A}(M)f^2 + M\mathbf{A}\left(f^2\right) = -\phi_{u_1}Mf^2 + 2\lambda_1 f^2 M = (2\lambda_1 - \phi_{u_1})Mf^2\\ &= (2\lambda_1 - \phi_{u_1})\zeta_1.\end{aligned}$$

Therefore ζ_1 is a solution of (35).

Conversely, let us suppose that λ_1 is a particular solution of (16) and that f, ζ_1 satisfy

$$\mathbf{A}(f) = \lambda_1 f, \qquad \mathbf{A}(\zeta_1) = (2\lambda_1 - \phi_{u_1})\zeta_1.$$

It can be checked that

$$M = \zeta_1 f^{-2} \tag{41}$$

is a Jacobi last multiplier for (1).

Similarly, it can be checked that if M is a Jacobi last multiplier for (1), λ_1 a particular solution of (16) and ζ_1 satisfies $\mathbf{A}(\zeta_1) = (2\lambda_1 - \phi_{u_1})\zeta_1$ then

$$f = \left(\frac{\zeta_1}{M}\right)^{1/2}$$

satisfies $\mathbf{A}(f) = \lambda_1 f$ and therefore $\mathbf{V} = f\partial_u + \lambda_1 f\partial_{u_1}$ is a commuting symmetry of (1).

Finally, let λ_1, λ_2 be two different particular solutions of (12), let f_1, f_2 be two non-null functions such that $\mathbf{A}(f_1) = \lambda_1 f_1$ and $\mathbf{A}(f_2) = \lambda_2 f_2$, respectively, and let

$W(f_1, f_2)$ be the function defined by (18). Then, by using (14), it can be directly checked that

$$M = (W(f_1, f_2))^{-1} = \left(\begin{vmatrix} 1 & u_1 & \phi \\ 0 & f_1 & \mathbf{A}(f_1) \\ 0 & f_2 & \mathbf{A}(f_2) \end{vmatrix} \right)^{-1} = \frac{1}{f_1\mathbf{A}(f_2) - f_2\mathbf{A}(f_1)} \tag{42}$$
$$= \frac{1}{f_1 f_2(\lambda_2 - \lambda_1)}$$

is a Jacobi last multiplier for (1). This result is, for second-order ordinary differential equations, a new version of a well-known result on the determination of a Jacobi last multiplier when a sufficient number (the order of the equation) of symmetries is known [12, 16].

4.2 Combining Solutions of (16)

The following theorem shows that a commuting symmetry and two different solutions of (16) let determine the remaining commuting symmetries.

Theorem 4 *Let $\mathbf{V}_1 = f_1(\partial_u + \lambda_1 \partial_{u_1})$ be a commuting symmetry of (1) and $\lambda_1 = \mathbf{A}(f_1)/f_1$. Let λ_2, λ_3 be two different solutions of (16).*

1. *The functions f_2 and f_3 defined by*

$$f_2 = f_1 \frac{\lambda_1 - \lambda_3}{\lambda_2 - \lambda_3} \quad \textit{and} \quad f_3 = f_1 \frac{\lambda_1 - \lambda_2}{\lambda_3 - \lambda_2} \tag{43}$$

   satisfy $\mathbf{A}(f_2) = \lambda_2 f_2$ and $\mathbf{A}(f_3) = \lambda_3 f_3$, respectively.
2. *If $\tilde{f}_2$ and $\tilde{f}_3$ are non-null functions such that $\mathbf{A}(\tilde{f}_2) = \lambda_2 \tilde{f}_2$ and $\mathbf{A}(\tilde{f}_3) = \lambda_3 \tilde{f}_3$ then the functions I_1 and I_2 defined by*

$$I_1 = \frac{f_1(\lambda_3 - \lambda_1)}{\tilde{f}_2(\lambda_3 - \lambda_2)}, \quad I_2 = \frac{f_1(\lambda_2 - \lambda_1)}{\tilde{f}_3(\lambda_2 - \lambda_3)} \tag{44}$$

   are first integrals of (1).

Proof By using Eq. (16), it can be checked that the functions h_2, h_3 defined by

$$h_2 = \frac{\lambda_1 - \lambda_3}{\lambda_2 - \lambda_3} \quad \text{and} \quad h_3 = \frac{\lambda_1 - \lambda_2}{\lambda_3 - \lambda_2} \tag{45}$$

satisfy

$$\mathbf{A}(h_2) = (\lambda_2 - \lambda_1)h_2, \qquad \mathbf{A}(h_3) = (\lambda_3 - \lambda_1)h_3. \tag{46}$$

Therefore, $f_2 = h_2 \cdot f_1$ and $f_3 = h_3 \cdot f_1$ satisfy

$$\mathbf{A}(f_2) = \mathbf{A}(h_2 f_1) = \lambda_2(h_2 f_1) = \lambda_2 f_2,$$
$$\mathbf{A}(f_3) = \mathbf{A}(h_3 f_1) = \lambda_3(h_3 f_1) = \lambda_2 f_3,$$

which proves the first part of this theorem.

The second part of this theorem can be directly proved because, for instance, the function I_1 given in (44) can be written as $I_1 = f_2/\tilde{f}_2$, being f_2 and $\tilde{f}_2$ solutions of the same linear equation $\mathbf{A}(f) = \lambda_2 f$. A similar situation happens with $I_2 = f_3/\tilde{f}_3$. Alternatively, this second part is a consequence of Theorem 2, because Eq. (44) is a consequence of (21) and the hypotheses of this Theorem.

Remark 5 Theorem 4 might be used in the following circumstance. Let us suppose that (∂_u, λ_2) and (∂_u, λ_3) are two $\mathcal{C}^\infty-$symmetries of (1) and let $\mathbf{v} = \xi\partial_x + \eta\partial_u$ be an additional known Lie point symmetry of (1) then, by setting $f_1 = Q$ and $\lambda_1 = \mathbf{A}(Q)/Q$ in (43) we obtain two functions f_2 and f_3 such that $\mathbf{V}_2 = f_2\left(\partial_u^{[\lambda_2,(1)]}\right)$ and $\mathbf{V}_3 = f_3\left(\partial_u^{[\lambda_3,(1)]}\right)$ are commuting symmetries of the given equation.

An example of the use of Remark 5 is given at the beginning of Example 3: a Lie point symmetry and two $\mathcal{C}^\infty-$symmetries can be used for the direct construction of a first integral of the equation. In our next example we consider the case where the equation has two known Lie point symmetries and a known $\mathcal{C}^\infty-$symmetry.

Example 2 In this example we consider the equation

$$u_2 + 4u^2u_1 + u^5 = 0. \tag{47}$$

This is the second equation in the Abel chain [10]. It is known that (47) admits the Lie point symmetries $\mathbf{v}_1$ and $\mathbf{v}_2$ given by

$$\mathbf{v}_1 = \partial_x, \qquad \mathbf{v}_2 = -2x\partial_x + u\partial_u. \tag{48}$$

The characteristics of these symmetries are $Q_1 = -u_1$ and $Q_2 = u + 2xu_1$, respectively. The corresponding $\mathcal{C}^\infty-$symmetries, written in canonical form, are (∂_u, λ_1) and (∂_u, λ_2), where λ_1 and λ_2 are given by

$$\lambda_1 = \frac{\mathbf{A}(Q_1)}{Q_1} = -4u^2 - \frac{u^5}{u_1}, \qquad \lambda_2 = \frac{\mathbf{A}(Q_2)}{Q_2} = \frac{(3 - 8xu^2)u_1 - 2xu^5}{u + 2xu_1}. \tag{49}$$

On the other hand, it has been proved in [10] that (47) admits the $\mathcal{C}^\infty-$symmetry

$$(\partial_u, \lambda_3) = \left(\partial_u, \frac{u_1}{u} - 2u^2\right), \tag{50}$$

which is also admitted by any equation in the Abel chain.

By using the first function in (43), taking $f_1 = Q_1$, we can directly obtain that the function

$$f_2 = Q_1 \frac{\lambda_1 - \lambda_3}{\lambda_2 - \lambda_3} = -\frac{(u_1 + u^3)(2xu_1 + u)}{2(xu_1 + xu^3 - u)} \tag{51}$$

satisfies $\mathbf{A}(f_2) = \lambda_2 f_2$. Similarly, by using the second function in (43) we obtain that

$$f_3 = Q_1 \frac{\lambda_1 - \lambda_2}{\lambda_3 - \lambda_2} = -\frac{(3u_1 + u^3)u}{xu_1 + xu^3 - u} \tag{52}$$

satisfies $\mathbf{A}(f_3) = \lambda_3 f_3$.

The functions in (44) let to obtain first integrals for Eq. (47); by considering the function $\tilde{f}_2 = Q_2$ we have that

$$I_1 = -\frac{u_1 + u^3}{2(xu_1 + xu^3 - u)} \tag{53}$$

is a non-trivial first integral of (47). If we take $\tilde{f}_3 = f_3$, then the second equation in (44) provides a constant first integral I_2.

In order to complete the integration of equation (47), we could use, for instance, the Jacobi last multiplier corresponding to (42):

$$M_{12} - \frac{1}{f_1 f_2(\lambda_2 - \lambda_1)} = -\frac{1}{(3u_1 + u^3)(u_1 + u^3)}.$$

The Jacobi last multiplier provides an integrating factor for the auxiliary equation $I_1 = C_1, C_1 \in \mathbb{R}$ written in normal form:

$$u_1 = \frac{-u(u^2(2C_1x + 1) - 2C_1)}{(2C_1x + 1)}.$$

Such integrating factor becomes (see [11, Theorem 8] and [6, Corollary 6] for details):

$$\mu = \frac{1}{2C_1u(u^2(2C_1x + 1) - 3C_1)}.$$

We omit here the expression of the general solution to equation (47) which can be obtained by using the classical method of Lie. The main purpose of this example is to illustrate how the combined use of two Lie point symmetries and one $\mathcal{C}^\infty$−symmetry provide directly a first integral and a Jacobi last multiplier which permit the complete integration of the equation.

4.3 First Integrals Derived from Several $\mathscr{C}^\infty$–Symmetries.

The following proposition shows that some first integrals of $\mathbf{A}$ can be obtained from solutions of several equations associated to the differential operator $\mathbf{A}$.

Proposition 4 *Let λ_1, λ_2 and λ_3 be three mutually different particular solutions of (12).*

1. *If ζ_1 satisfies $\mathbf{A}(\zeta_1) = (2\lambda_1 - \phi_{u_1})\zeta_1$ then*

$$I_{123} = \frac{1}{\zeta_1}\left(\frac{1}{\lambda_3 - \lambda_1} - \frac{1}{\lambda_2 - \lambda_1}\right) \tag{54}$$

is a first integral of $\mathbf{A}$.

2. *If h is a non-null function satisfying $\mathbf{A}(h) = (\lambda_2 - \lambda_1)h$ then*

$$\widetilde{I}_{123} = \frac{\lambda_1 - \lambda_3}{h(\lambda_3 - \lambda_2)} \tag{55}$$

is a first integral of $\mathbf{A}$.

Proof (1) By Theorem 3 the function λ_3 can be written in the form $\lambda_3 = \frac{1}{1+h_0}\lambda_1 + \frac{h_0}{1+h_0}\lambda_2$, where $h_0 = \frac{\lambda_1-\lambda_3}{\lambda_3-\lambda_2}$ satisfies $\mathbf{A}(h_0) = (\lambda_2 - \lambda_1)h_0$. On the other hand, (38)–(39) implies that h_0 must be of the form $h_0 = 1/\big((\lambda_2 - \lambda_1)I_{123}\zeta_1\big)$, where I_{123} is a first integral of $\mathbf{A}$. Now (54) is a consequence of

$$\frac{\lambda_1 - \lambda_3}{\lambda_3 - \lambda_2} = \frac{1}{(\lambda_2 - \lambda_1)I_{123}\zeta_1}.$$

(2) Since h and h_0 are solutions of the equation $\mathbf{A}(h) = (\lambda_2 - \lambda_1)h$, necessarily there exist a first integral $\widetilde{I}_{123}$ such that $h_0 = \widetilde{I}_{123}h$. This implies (55).

When four particular solutions of the determining equation (12) are known, the ratio of the corresponding first integrals I_{123} and I_{124} is also a first integral of $\mathbf{A}$.

Corollary 1 *Let $\lambda_1, \lambda_2, \lambda_3$ and λ_4 be particular solutions of (12). Then*

$$I = \frac{(\lambda_2 - \lambda_4)(\lambda_3 - \lambda_1)}{(\lambda_2 - \lambda_3)(\lambda_4 - \lambda_1)} \tag{56}$$

is a first integral of $\mathbf{A}$.

Proof The function I given in (56) is the ratio of the corresponding first integrals I_{123} and I_{124} given in (54), when the same solution ζ_1 of $\mathbf{A}(\zeta_1) = (2\lambda_1 - \phi_{u_1})\zeta_1$ is used for both first integrals. Alternatively, if h is a solution of the equation $\mathbf{A}(h) = (\lambda_2 - \lambda_1)h$ and $\widetilde{I}_{123}$ and $\widetilde{I}_{124}$ are the corresponding first integrals given by (55), then I is also the ratio of $\widetilde{I}_{123}$ and $\widetilde{I}_{124}$.

Example 3 In this example we consider, again, the Morse equation (30) to illustrate how the solutions of several equations related to the vector field **A** can be combined to obtain different objects related to the given equation.

Morse equation admits the trivial Lie point symmetry ∂_x. Its characteristic is $f_1 = Q_1 = -u_1$ and

$$\lambda_1 = \frac{\mathbf{A}(f_1)}{f_1} = \frac{e^{-u} - 1 - u_1^2}{u_1}$$

defines a $\mathscr{C}^\infty$−symmetry (∂_u, λ_1) for (30). By using the $\mathscr{C}^\infty$−symmetries

$$(\partial_u, \lambda_2) = (\partial_u, -u_1 + \cot x), \qquad (\partial_u, \lambda_3) = (\partial_u, -u_1 - \tan x),$$

which were defined in (31), and Remark 5, we can obtain without any kind of integration two functions f_2 and f_3 such that $\mathbf{A}(f_2) = \lambda_2 f_2$ and $\mathbf{A}(f_3) = \lambda_3 f_3$, respectively. By using (43) we get that

$$f_2 = \sin x \left(-\sin x \cdot u_1 - \cos x \cdot (\mathrm{e}^{-u} - 1)\right) \tag{57}$$

and

$$f_3 = \cos x \left(-\cos x \cdot u_1 + \sin x \cdot (\mathrm{e}^{-u} - 1)\right), \tag{58}$$

determine two functions that define, respectively, two commuting symmetries $\mathbf{V}_2$ and $\mathbf{V}_3$ such that $\mathbf{A}, \mathbf{V}_2, \mathbf{V}_3$ are linearly independent.

It can be checked that $h_0 = \frac{g_2(x)}{g_1(x)} = \cot x$ satisfies $\mathbf{A}(h_0) = (\lambda_3 - \lambda_2)h_0$. Furthermore,

$$h_1 = \frac{f_3}{f_2} = \frac{\cos x \left(-\cos x \cdot u_1 + (e^{-u} - 1)\sin x\right)}{\sin x\,(-\sin x \cdot u_1 - (e^{-u} - 1)\cos x)}$$

satisfies $\mathbf{A}(h_1) = (\lambda_3 - \lambda_2)h_1$. Therefore h_0 and h_1 are solutions of the same equation $\mathbf{A}(h) = (\lambda_3 - \lambda_2)h$. Consequently, $J_0 = \frac{h_1}{h_0}$ is a first integral of (30):

$$J_0 = \frac{h_1}{h_0} = \frac{-\cos x \cdot u_1 + (e^{-u} - 1)\sin x}{-\sin x \cdot u_1 - (e^{-u} - 1)\cos x}. \tag{59}$$

On the other hand, if we use h_0 instead of h in (55) we get that the first integral

$$J_1 = \frac{1}{h_0}\frac{\lambda_2 - \lambda_1}{\lambda_1 - \lambda_3}$$

is J_0. Of course, we could also have used h_1 instead of h in (55); however, it can be checked that the first integral calculated in this way is constant.

So far, the only Lie point symmetry we have used in this example is the trivial symmetry $\mathbf{v}_1 = \partial_x$. Let us recall that the equations in class $\mathscr{A}_1$ admit a maximal algebra of Lie point symmetries [8]. If $\mathbf{v}_1 = \partial_x, \mathbf{v}_2, \ldots, \mathbf{v}_8$ are the generators of this Lie algebra and the $f_i = Q_i$, $1 \le i \le 8$, are the corresponding characteristics then the functions $\lambda_i = \mathbf{A}(Q_i)/Q_i$, are particular solutions of the corresponding determining equation (12) (or (16)). These functions could be used to compute first integrals of the equation by using Proposition 4 or Corollary 1, by using λ_i and Q_i instead of λ_1 and Q_1, for $2 \leq i \leq 8$. Of course, some of the first integrals that can be calculated in this way could be functionally dependent on the first integral previously obtained.

Example 4 The family of 2nd-order equations

$$u_2 = \frac{1}{2u}u_1^2 - 2uu_1 - \frac{1}{2}u^3 + ku - \frac{1}{2u}, \qquad k \in \mathbb{R} \tag{60}$$

is a particular case of the XXVII equation in the Painlevé-Gambier classification [5], which clearly admits $\mathbf{v}_1 = \partial_x$ as Lie point symmetry. It can be checked that $\mathbf{v}$ is the unique Lie point symmetry admitted by (60). According to Proposition 2, $\mathbf{v}_1$ defines the commuting symmetry $\mathbf{V}_1 = Q_1\partial_u + \mathbf{A}(Q_1)\partial_{u_1}$, where $Q_1 = -u_1$ is the characteristic of $\mathbf{v}_1$ and $\mathbf{A}$ denotes the vector field associated to (60).

It is also known [3] that Eq. (60) admits the $\mathscr{C}^\infty-$symmetries (∂_u, λ_2) and (∂_u, λ_3), defined by:

$$\lambda_2 = \frac{u_1}{u} - u + \frac{1}{u}, \quad \lambda_3 = \frac{u_1}{u} - u - \frac{1}{u}. \tag{61}$$

By using (43), the functions

$$\begin{aligned} f_2 &= Q_1\frac{\lambda_1 - \lambda_3}{\lambda_2 - \lambda_3} = \frac{1}{4}\left((u_1 - 1)^2 + u^2(2u_1 - 2k + u^2)\right), \\ f_3 &= Q_1\frac{\lambda_1 - \lambda_2}{\lambda_3 - \lambda_2} = \frac{1}{4}\left(-(u_1 + 1)^2 - u^2(2u_1 - 2k + u^2)\right) \end{aligned} \tag{62}$$

satisfy $\mathbf{A}(f_2) = \lambda_2 f_2$ and $\mathbf{A}(f_3) = \lambda_3 f_3$, respectively. For this example, it can be checked that the corresponding first integrals that appear in (44) are constant and hence they are not useful to reduce or integrate equation (60).

Next, we show how the functions given in (61) and (62) can still be used to integrate equation (60) by quadratures. By Theorem 1, for $i = 2, 3$, the vector field $\mathbf{V}_i = f_i\partial_u + \lambda_i f_i \partial_{u_1}$ is a commuting symmetry of (60) which is $\mathbf{A}-$equivalent to the $\mathscr{C}^\infty-$symmetry (∂_u, λ_i). Besides $[\mathbf{V}_2, \mathbf{V}_3] = 0$ and by Theorem 5 in [11] two functionally independent first integrals can be computed by quadratures from systems:

$$(I_1)_x = \frac{\lambda_2 u_1 - \phi}{f_3(\lambda_3 - \lambda_2)}, \quad (I_1)_u = \frac{-\lambda_2}{f_3(\lambda_3 - \lambda_2)}, \quad (I_1)_{u_1} = \frac{1}{f_3(\lambda_3 - \lambda_2)}, \tag{63}$$

and

$$(I_2)_x = \frac{\lambda_3 u_1 - \phi}{f_2(\lambda_2 - \lambda_3)}, \quad (I_2)_u = \frac{-\lambda_3}{f_2(\lambda_2 - \lambda_3)}, \quad (I_2)_{u_1} = \frac{1}{f_2(\lambda_2 - \lambda_3)}, \tag{64}$$

where ϕ denotes the right-hand side of (60). The corresponding first integrals become

$$I_1 = \begin{cases} x - \sqrt{\dfrac{2}{1-k}}\ \arctan\left(\dfrac{u_1 + u^2 - 1}{u\,\sqrt{2(1-k)}}\right), & \text{for } k < 1 \\ -x + \dfrac{2\,u}{u_1 + u^2 - 1}, & \text{for } k = 1 \\ x - \sqrt{\dfrac{2}{k+1}}\ \operatorname{arctanh}\left(\dfrac{u_1 + u^2 + 1}{u\,\sqrt{2(k+1)}}\right), & \text{for } k > 1. \end{cases}$$

and

$$I_2 = \begin{cases} x - \sqrt{\dfrac{2}{-(k+1)}}\ \arctan\left(\dfrac{u_1 + u^2 + 1}{u\,\sqrt{-2(k+1)}}\right), & \text{for } k < -1 \\ x - \dfrac{2\,u}{u_1 + u^2 + 1}, & \text{for } k = -1 \\ x - \sqrt{\dfrac{2}{k+1}}\ \operatorname{arctanh}\left(\dfrac{u_1 + u^2 + 1}{u\,\sqrt{2(k+1)}}\right), & \text{for } k > -1. \end{cases}$$

Once the complete set $\{I_1, I_2\}$ of first integrals has been determined, the general solution to equation (60) can be easily obtained by eliminating u_1 from the system $I_1 = C_1, I_2 = C_2$, where $C_1, C_2 \in \mathbb{R}$.

5 Concluding Remark

Several new interconnections between the determining equations for generalized, commuting, and generalized $\mathscr{C}^\infty$−symmetries, as well as other equations related to the linear operator associated to a second-order ODE, have been established. This lets to exploit and combine various solutions to some of these equations in order to find particular solutions of some of the others. As a consequence, new ways of computing first integrals and Jacobi last multipliers of the equation have been found; these quantities play a fundamental role in the integrability of the equations under study.

Acknowledgements The authors acknowledge the financial support from the University of Cádiz by means of the project PR2017-090 and from the *Junta de Andalucía* research group FQM-377.

A. Ruiz acknowledges the financial support from the *Ministerio de Educación, Cultura y Deporte* of Spain by means of a FPU grant (FPU15/02872).

References

1. Bluman, G.W., Kumei, S.: Symmetries and Differential Equations. Springer, New York (1989)
2. Duarte, L.G.S., Moreira, I.C., Santos, F.C.: Linearization under non-point transformations. J. Phys. A Math. Gen. **27**, L739–L743 (1994)
3. Guha, P., Choudhury, A., Khanra, B.: λ−Symmetries, isochronicity and integrating factors of nonlinear ordinary differential equations. J. Eng. Math. **82**(1), 85–99 (2013). https://doi.org/10.1007/s10665-012-9614-5

4. Ibragimov, N.H.: A Practical Course in Differential Equations and MathematicalModelling: Classical and New Methods, Nonlinear Mathematical Models Symmetry and Invariance Principles. World Scientific, Beijing (2010)
5. Ince, E.: Ordinary Differential Equations. Dover, New York (1956)
6. Muriel, C., Romero, J.: The λ-symmetry reduction method and Jacobi last multipliers. Commun. Nonlinear Sci. Numer. Simul. **19**(4), 807–820 (2014)
7. Muriel, C., Romero, J.L.: New methods of reduction for ordinary differential equations. IMA J. Appl. Math. **66**(2), 111–125 (2001). https://doi.org/10.1093/imamat/66.2.111
8. Muriel, C., Romero, J.L.: Second-order ordinary differential equations and first integrals of the form $A(t,x)\dot{x} + B(t,x)$. J. Nonlinear Math. Phys. **16**(1), 209–222 (2009). https://doi.org/10.1142/S1402925109000418
9. Muriel, C., Romero, J.L.: Nonlocal transformations and linearization of second-order ordinary differential equations. J. Phys. A Math. Theor. **43**(43), 434,025 (13 pp) (2010). https://doi.org/10.1088/1751-8113/43/43/434025
10. Muriel, C., Romero, J.L.: λ-symmetries of some chains of ordinary differential equations. Nonlinear Anal. Real World Appl. **16**, 191–201 (2014)
11. Muriel, C., Romero, J.L., Ruiz, A.: λ-symmetries and integrability by quadratures. IMA J. Appl. Math. **82**(5), 1061–1087 (2017). https://doi.org/10.1093/imamat/hxx024
12. Nucci, M.: Jacobi last multiplier and Lie symmetries: a novel application of an old relationship. J. Nonlinear Math. Phys. **12**(2), 284–304 (2005)
13. Olver, P.J.: Applications of Lie Groups to Differential Equations. Graduate Texts in Mathematics, vol. 107. Springer, New York (1986). https://doi.org/10.1007/978-1-4684-0274-2
14. Ovsiannikov, L.V.: Group Analysis of Differential Equations. Academic, New York (1982). [Harcourt Brace Jovanovich Publishers]. Translated from the Russian by Y. Chapovsky, Translation edited by William F. Ames
15. Stephani, H.: Differential Equations: Their Solution Using Symmetries. Cambridge University, Cambridge (1989)
16. Whittaker, E., McCrae, W.: A Treatise on the Analytical Dynamics of Particles and Rigid Bodies. Cambridge Mathematical Library. Cambridge University, Cambridge (1988)

Differential Invariants for Two and Three Dimensional Linear Parabolic Equations

Adnan Aslam, Asghar Qadir and Muhammad Safdar

Abstract We find equivalence transformations for linear parabolic equations having two and three spatial dimensions. Invariants associated with these higher dimensional linear parabolic equations are derived using the obtained set of equivalence transformations. We apply Lie infinitesimal method to deduce the associated invariants. We find first order invariants for the the higher dimensional parabolic equations due to an invertible change of the dependent and independent variables separately. Further, obtained invariants are employed to reduce these linear higher dimensional parabolic equations to their simplest forms.

Keywords Semi-invariants · Joint differential invariants · Lie infinitesimal method

1 Introduction

Lie infinitesimal method has been widely employed to derive differential invariants for linear and nonlinear differential equations (DEs). The said method comprises of two steps: Derivation of the equivalence group associated with the concerned equation and calculation of invariants using this equivalence group. Equivalence transformations are invertible transformations of the dependent and independent

A. Aslam (✉)
Department of Basic Sciences, School of Electrical Engineering and Computer Science (SEECS), National University of Sciences and Technology (NUST), Islamabad H-12, Pakistan
e-mail: adnan.aslam@seecs.edu.pk

A. Qadir
School of Natural Sciences (SNS), National University of Sciences and Technology (NUST), Islamabad H-12, Pakistan
e-mail: aqadirmath@yahoo.com

M. Safdar
School of Mechanical and Manufacturing Engineering (SMME), National University of Sciences and Technology (NUST), Islamabad H-12, Pakistan
e-mail: safdar.camp@gmail.com

V. G. Kac et al. (eds.), *Symmetries, Differential Equations and Applications*, Springer Proceedings in Mathematics & Statistics 266,
https://doi.org/10.1007/978-3-030-01376-9_9

variables that preserve e.g. the order, form, linearity and homogeneity of a DE . Basic theory of equivalence groups, differential invariants and algorithms to derive them can be found in [1–3]. These invariants ensure existence of point transformations, i.e., an invertible change of the dependent and/or independent variables to map two equations into each other, if for both the equations associated invariants remain unaltered.

Developing integration theory for hyperbolic equations

$$u_{xy} + a(x, y)u_x + b(x, y)u_y + c(x, y)u = 0, \tag{1}$$

Laplace [4] found the following quantities

$$\lambda_1 := a_x + ab - c, \quad \lambda_2 := b_y + ab - c, \tag{2}$$

that remain unaltered (called semi-invariants) when *only* dependent variable undergoes a linear change $u = \sigma(x, y)\bar{u}$. These are first order semi-invariants as they contain at most first order partial derivatives of the coefficients $a(x, y)$, and $b(x, y)$. Further, Cotton [5] investigated Laplace invariants of the elliptic equation

$$u_{xx} + u_{yy} + a(x, y)u_x + b(x, y)u_y + c(x, y)u = 0, \tag{3}$$

and he came across the following first order invariants

$$\lambda_3 := a_y - b_x, \quad \lambda_4 := a_x + b_y + \frac{1}{2}(a^2 + b^2) - 2c. \tag{4}$$

Ibragimov in his attempt to fill the gap by finding invariants of the parabolic equation

$$u_t + a(t, x)u_{xx} + b(t, x)u_x + c(t, x)u = 0, \tag{5}$$

do not only derived the associated invariant

$$\lambda_5 := \frac{1}{2}b^2 a_x + (a_t + aa_{xx} - a_x^2)b + (aa_x - ab)b_x - ab_t - a^2 b_{xx} + 2a^2 c_x, \tag{6}$$

for parabolic equation by employing the infinitesimal approach, indeed he re-derived semi-invariants (2) and (4) using the said method [6], for the hyperbolic and elliptic equations respectively. Attempts have also been made to find semi-invariants associated with the partial differential equations (PDEs) of the types mentioned, due to a change of the independent variables. Joint invariants of these equations via an infinitesimal change of both the dependent and independent variables are also obtained [7–10]. Further, infinitesimal approach has also been engaged to find invariants of the linear systems of two elliptic, hyperbolic and parabolic type equations [11–14]. Moreover, by means of Lie infinitesimal method invariants for two and three-dimensional hyperbolic equations have been deduced [15].

In this paper we obtain semi-invariants associated with the parabolic equations

$$E_1 : u_t = a_1(t,x,y)u_{xx} + a_2(t,x,y)u_{yy} + b_1(t,x,y)u_x + b_2(t,x,y)u_y + c(t,x,y)u, \quad (7)$$

and

$$E_2 : u_t = a_1(t,x,y,z)u_{xx} + a_2(t,x,y,z)u_{yy} + a_3(t,x,y,z)u_{zz} + b_1(t,x,y,z)u_x + b_2(t,x,y,z)u_y + b_3(t,x,y,z)u_z + c(t,x,y,z)u, \quad (8)$$

where the subscripts denote partial derivatives, due to a change of the dependent and independent variables, separately. Firstly we find the associated group of equivalence transformations for both (7) and (8). Secondly we derive semi-invariants associated with these equations by adopting infinitesimal method. These invariants are obtained by considering infinitesimal transformations of both the dependent and independent variables, separately. We show that with the help of the deduced invariants Eqs. (7) and (8) are reducible to

$$u_t = a_1(t,x,y)u_{xx} + a_2(t,x,y)u_{yy}, \quad (9)$$

and

$$u_t = a_1(t,x,y,z)u_{xx} + a_2(t,x,y,z)u_{yy} + a_3(t,x,y,z)u_{zz}, \quad (10)$$

respectively, under a linear transformation of the dependent variables. Further, due to an invertible change of the independent variables Eqs. (7) and (8) are transformable to the following simple forms

$$u_t = \alpha_1 u_{xx} + \alpha_2(x)u_{yy} + c(t,x,y)u, \quad (11)$$

and

$$u_t = \alpha_1 u_{xx} + \alpha_2(x)u_{yy} + \alpha_3 u_{zz} + c(t,x,y,z)u, \quad (12)$$

respectively, where α_1 and α_3, are nonzero constants, $\alpha_{2,x} \neq 0$ and c remains the same as is in (7) and (8).

In the second section equivalence transformations for Eqs. (7) and (8) are determined. The third section is on the derivation of the semi-invariants associated with these equations. Algorithm to find point transformations to map higher dimensional parabolic equations into each other is presented in fourth section. Subsequent section is on application of the obtained invariants which when agree for two higher-dimensional linear parabolic equations are shown to map them into each other. The last section concludes our work.

2 Equivalence Transformation Groups for Two and Three-Dimensional Parabolic Equations

Equivalence transformations play an essential role in deriving invariants of differential equations using infinitesimal method. The group formed by set of all equivalence transformations associated with a family of equations is called equivalence group. There are two methods to obtain these transformations [3, 16], the first one is called direct method while the second one is known as Lie infinitesimal method. Although the direct method has the benefit of providing the most general group of equivalence transformations, it is seldom used due to considerable computational difficulties involved. Here we use the infinitesimal method to get the equivalence transformations for Eqs. (7) and (8).

Consider the parabolic equation (7) with three independent variables to derive the equivalence transformations which map this family into itself, in general with different coefficients e.g. $\overline{a_j}$, $\overline{b_j}$ and $\overline{c}$, where $j = 1, 2$. These transformations in general form are

$$\bar{t} = \phi_1(t, x, y, u),\ \bar{x} = \phi_2(t, x, y, u),\ \bar{y} = \phi_3(t, x, y, u),\ \bar{u} = \psi(t, x, y, u). \quad (13)$$

The arbitrary functions ϕ_k and ψ for $k = 1, 2, 3$, are such that $\partial(\phi_k, \psi)/\partial(t, x, y, u) \neq 0$. An operator of the form

$$\begin{aligned} \mathbf{X} = \ & \xi_1\frac{\partial}{\partial t} + \xi_2\frac{\partial}{\partial x} + \xi_3\frac{\partial}{\partial y} + \eta\frac{\partial}{\partial u} + \mu_1\frac{\partial}{\partial a_1} + \mu_2\frac{\partial}{\partial a_2} \\ & + \mu_3\frac{\partial}{\partial b_1} + \mu_4\frac{\partial}{\partial b_2} + \mu_5\frac{\partial}{\partial c}, \end{aligned} \quad (14)$$

where $\xi_k = \xi_k(t, x, y, u)$ for $k = 1, 2, 3$, $\eta = \eta(t, x, y, u)$ and $\mu_m = \mu_m(t, x, y, u, a_j, b_j, c)$ for $m = 1, 2, 3, 4, 5$, and $j = 1, 2$, is used to derive the continuous group of equivalence transformations associated with Eq. (7). We employ the twice extended generator (14) on (7), that is

$$\mathbf{X}^{[2]}(E_1)|_{(E_1)} = 0. \quad (15)$$

The notation $|_{E_{(1)}}$ means evaluated on Eq. (7) and

$$\mathbf{X}^{[2]} = \mathbf{X} + \eta^t\frac{\partial}{\partial u_t} + \eta^x\frac{\partial}{\partial u_x} + \eta^y\frac{\partial}{\partial u_y} + \eta^{xx}\frac{\partial}{\partial u_{xx}} + \eta^{yy}\frac{\partial}{\partial u_{yy}}, \quad (16)$$

where

$$\begin{aligned} \eta^t &= D_t\eta - u_t D_t\xi_1 - u_x D_t\xi_2 - u_y D_t\xi_3, \\ \eta^x &= D_x\eta - u_t D_x\xi_1 - u_x D_x\xi_2 - u_y D_x\xi_3, \\ \eta^y &= D_y\eta - u_t D_y\xi_1 - u_x D_y\xi_2 - u_y D_y\xi_3, \end{aligned}$$

$$\eta^{xx} = D_x\eta^x - u_{tx}D_x\xi_1 - u_{xx}D_x\xi_2 - u_{xy}D_x\xi_3,$$
$$\eta^{yy} = D_y\eta^y - u_{ty}D_y\xi_1 - u_{xy}D_y\xi_2 - u_{yy}D_y\xi_3. \tag{17}$$

The operators D_t, D_x, and D_y denote the total derivatives with respect to $t,\ x$, and y, respectively

$$D_t = \frac{\partial}{\partial t} + u_t\frac{\partial}{\partial u} + u_{tt}\frac{\partial}{\partial u_t} + u_{tx}\frac{\partial}{\partial u_x} + u_{ty}\frac{\partial}{\partial u_y} + \cdots,$$
$$D_x = \frac{\partial}{\partial x} + u_x\frac{\partial}{\partial u} + u_{tx}\frac{\partial}{\partial u_t} + u_{xx}\frac{\partial}{\partial u_x} + u_{xy}\frac{\partial}{\partial u_y} + \cdots,$$
$$D_y = \frac{\partial}{\partial y} + u_y\frac{\partial}{\partial u} + u_{ty}\frac{\partial}{\partial u_t} + u_{xy}\frac{\partial}{\partial u_x} + u_{yy}\frac{\partial}{\partial u_y}\cdots. \tag{18}$$

For detailed extension procedure reader is referred to [3, 17]. All the coefficients of derivatives of $u(t, x, y)$ and their different powers readable from (15), provide a set of determining equations when they all are set equal to zero. These determining equations constitute a system of linear PDEs. To solve this system we used MAPLE which reveals the infinitesimal coordinates of (14) that read as

$$\begin{aligned}
\xi_1 &= \xi_1(t),\\
\xi_2 &= \xi_2(t, x),\\
\xi_3 &= \xi_3(t, y),\\
\eta &= \eta(t, x, y)u,\\
\mu_1 &= a_1(2\xi_{2_x} - \xi_{1_t}),\\
\mu_2 &= a_2(2\xi_{3_y} - \xi_{1_t}),\\
\mu_3 &= a_1(\xi_{2_{xx}} - 2\eta_x) + b_1(\xi_{2_x} - \xi_{1_t}) - \xi_{2_t},\\
\mu_4 &= a_2(\xi_{3_{yy}} - 2\eta_y) + b_2(\xi_{3_y} - \xi_{1_t}) - \xi_{3_t},\\
\mu_5 &= \eta_t - a_1\eta_{xx} - a_2\eta_{yy} - b_1\eta_x - b_2\eta_y - c\xi_{1_t},
\end{aligned} \tag{19}$$

where $\xi_1(t)$, $\xi_2(t, x)$, $\xi_3(t, y)$ and $\eta(t, x, y)$ are arbitrary functions and t, x, y in the subscripts denote their partial derivatives. We apply (14) with the infinitesimal coordinates (19) in the next section to derive semi-invariants associated with the Eq. (7). We construct semi-invariants for the class (7), considering first a linear change of the dependent variable

$$\bar{u} = \psi(t, x, y, u) := \eta(t, x, y)u, \tag{20}$$

and then an arbitrary change of the independent variables

$$\bar{t} = \phi_1(t, x, y, u) := \xi_1(t),$$
$$\bar{x} = \phi_2(t, x, y, u) := \xi_2(t, x),$$

$$\overline{y} = \phi_3(t, x, y, u) := \xi_3(t, y). \tag{21}$$

For the second class of parabolic equations (8) we follow the same procedure to find equivalence group of transformations. Here the operator takes the following form

$$\begin{aligned}\mathbf{Z} = &\ \xi_1\frac{\partial}{\partial t} + \xi_2\frac{\partial}{\partial x} + \xi_3\frac{\partial}{\partial y} + \xi_4\frac{\partial}{\partial z} + \eta\frac{\partial}{\partial u} + \mu_1\frac{\partial}{\partial a_1} + \mu_2\frac{\partial}{\partial a_2} + \mu_3\frac{\partial}{\partial a_3}\\ &+\mu_4\frac{\partial}{\partial b_1} + \mu_5\frac{\partial}{\partial b_2} + \mu_6\frac{\partial}{\partial b_3} + \mu_7\frac{\partial}{\partial c},\end{aligned} \tag{22}$$

where $\xi_l = \xi_l(t, x, y, z, u)$ for $l = 1, 2, 3, 4$, $\eta = \eta(t, x, y, z, u)$ and $\mu_p = \mu_p(t, x, y, z, u, a_k, b_k, c)$ for $p = 1, 2, 3, 4, 5, 6, 7$, and $k = 1, 2, 3$, to get the required equivalence mappings. Inserting the set of determining equations obtained from

$$\mathbf{Z}^{[2]}(E_2)|_{(E_2)} = 0, \tag{23}$$

in MAPLE provides the following infinitesimal coordinates

$$\begin{aligned}
\xi_1 &= \xi_1(t),\\
\xi_2 &= \xi_2(t, x),\\
\xi_3 &= \xi_3(t, y),\\
\xi_4 &= \xi_4(t, z),\\
\eta &= \eta(t, x, y, z)u,\\
\mu_1 &= a_1(2\xi_{2_x} - \xi_{1_t}),\\
\mu_2 &= a_2(2\xi_{3_y} - \xi_{1_t}),\\
\mu_3 &= a_3(2\xi_{4_z} - \xi_{1_t}),\\
\mu_4 &= a_1(\xi_{2_{xx}} - 2\eta_x) + b_1(\xi_{2_x} - \xi_{1_t}) - \xi_{2_t},\\
\mu_5 &= a_2(\xi_{3_{yy}} - 2\eta_y) + b_2(\xi_{3_y} - \xi_{1_t}) - \xi_{3_t},\\
\mu_6 &= a_3(\xi_{4_{zz}} - 2\eta_z) + b_3(\xi_{4_z} - \xi_{1_t}) - \xi_{4_t},\\
\mu_7 &= \eta_t - a_1\eta_{xx} - a_2\eta_{yy} - a_3\eta_{zz} - b_1\eta_x - b_2\eta_y - b_3\eta_z - c\xi_{1_t},
\end{aligned} \tag{24}$$

for (22). The operator of the equivalence transformations (22) with the above infinitesimal coordinates is employed in subsequent section to derive semi-invariants for (8). We consider transformations of the dependent and independent variables provided above, separately, to derive associated semi-invariants.

A zeroth order differential invariant of a PDE under equivalence group of transformations is a function of all or a few coefficients of the concerned equation. For instance, in case of (7) and (8) it reads as

$$J_0(a_j, b_j, c), \quad j = 1, 2, \tag{25}$$

and

$$J_0(a_k, b_k, c), \quad k = 1, 2, 3, \tag{26}$$

respectively. Similarly, a first order differential invariant associated with both the PDEs is

$$J_1(a_j, b_j, c, a_{j,\sigma}, b_{j,\sigma}, c_\sigma), \quad \sigma \in \{t, x, y\}, \tag{27}$$

and

$$J_1(a_k, b_k, c, a_{k,\tau}, b_{k,\tau}, c_\tau), \quad \tau \in \{t, x, y, z\}, \tag{28}$$

respectively. Notice that J_1 contains first order partial derivatives of the coefficients of (7) and (8). Likewise, a differential invariant of an arbitrary order q involves qth order derivatives of all or a few coefficients of the concerned equation.

To find an invariant of order zero for a PDE we incorporate operators $\mathbf{X}_k$, for $k = 1, 2, \ldots, n$, in the invariance criterion

$$\mathbf{X}_k J_0 = 0, \tag{29}$$

where $\mathbf{X}_1$,$\mathbf{X}_2$, ...,$\mathbf{X}_n$, characterize continuous group of equivalence transformations associated with the concerned equation. If a generator $\mathbf{X}_p$, contains l arbitrary functions and m derivatives of these functions then $l + m$, is the number of linear PDEs obtained by the invariance criterion. In order to find the first and higher order differential invariants of a PDE one needs to extend the operators $\mathbf{X}_k$, $k = 1, 2, \ldots, n$,. Such an operator for (7) is given by (14), which can be extended as

$$\begin{aligned}
\mathbf{X}^{[1]} &= \mathbf{X} + \mu_1^{i_1}\frac{\partial}{\partial a_{1,i_1}} + \mu_2^{i_1}\frac{\partial}{\partial a_{2,i_1}} + \mu_3^{i_1}\frac{\partial}{\partial b_{1,i_1}} + \mu_4^{i_1}\frac{\partial}{\partial b_{2,i_1}} + \mu_5^{i_1}\frac{\partial}{\partial c_{i_1}},\\
\mathbf{X}^{[2]} &= \mathbf{X}^{[1]} + \mu_1^{i_1i_2}\frac{\partial}{\partial a_{1,i_1i_2}} + \mu_2^{i_1i_2}\frac{\partial}{\partial a_{2,i_1i_2}} + \mu_3^{i_1i_2}\frac{\partial}{\partial b_{1,i_1i_2}} + \mu_4^{i_1i_2}\frac{\partial}{\partial b_{2,i_1i_2}} + \mu_5^{i_1i_2}\frac{\partial}{\partial c_{i_1i_2}},\\
&\vdots \qquad\qquad \vdots \qquad\qquad \vdots\\
\mathbf{X}^{[q]} &= \mathbf{X}^{[q-1]} + \mu_1^{i_1i_2\cdots i_q}\frac{\partial}{\partial a_{1,i_1i_2\cdots i_q}} + \mu_2^{i_1i_2\cdots i_q}\frac{\partial}{\partial a_{2,i_1i_2\cdots i_q}} + \mu_3^{i_1i_2\cdots i_q}\frac{\partial}{\partial b_{1,i_1i_2\cdots i_q}}\\
&\quad + \mu_4^{i_1i_2\cdots i_q}\frac{\partial}{\partial b_{2,i_1i_2\cdots i_q}} + \mu_5^{i_1i_2\cdots i_q}\frac{\partial}{\partial c_{2,i_1i_2\cdots i_q}}, \quad i_1, i_2, \ldots, i_q \in \{t, x, y\},
\end{aligned} \tag{30}$$

where

$$\begin{aligned}
\mu_1^{i_1i_2\cdots i_q} &= \overline{D}_{i_q}\mu_1^{i_1i_2\cdots i_{q-1}} - a_{1,i_1i_2\cdots i_{q-1}t}\overline{D}_{i_q}(\xi_1) - a_{1,i_1i_2\cdots i_{q-1}x}\overline{D}_{i_q}(\xi_2) - a_{1,i_1i_2\cdots i_{q-1}y}\overline{D}_{i_q}(\xi_3),\\
\mu_2^{i_1i_2\cdots i_q} &= \overline{D}_{i_q}\mu_2^{i_1i_2\cdots i_{q-1}} - a_{2,i_1i_2\cdots i_{q-1}t}\overline{D}_{i_q}(\xi_1) - a_{2,i_1i_2\cdots i_{q-1}x}\overline{D}_{i_q}(\xi_2) - a_{2,i_1i_2\cdots i_{q-1}y}\overline{D}_{i_q}(\xi_3),
\end{aligned}$$

$$
\begin{aligned}
\mu_3^{i_1 i_2 \cdots i_q} &= \overline{D}_{i_q}\mu_3^{i_1 i_2 \cdots i_{q-1}} - b_{1,i_1 i_2 \cdots i_{q-1}t}\overline{D}_{i_q}(\xi_1) - b_{1,i_1 i_2 \cdots i_{q-1}x}\overline{D}_{i_q}(\xi_2) - b_{1,i_1 i_2 \cdots i_{q-1}y}\overline{D}_{i_q}(\xi_3),\\
\mu_4^{i_1 i_2 \cdots i_q} &= \overline{D}_{i_q}\mu_4^{i_1 i_2 \cdots i_{q-1}} - b_{2,i_1 i_2 \cdots i_{q-1}t}\overline{D}_{i_q}(\xi_1) - b_{2,i_1 i_2 \cdots i_{q-1}x}\overline{D}_{i_q}(\xi_2) - b_{2,i_1 i_2 \cdots i_{q-1}y}\overline{D}_{i_q}(\xi_3),\\
\mu_5^{i_1 i_2 \cdots i_q} &= \overline{D}_{i_q}\mu_5^{i_1 i_2 \cdots i_{q-1}} - c_{i_1 i_2 \cdots i_{q-1}t}\overline{D}_{i_q}(\xi_1) - c_{i_1 i_2 \cdots i_{q-1}x}\overline{D}_{i_q}(\xi_2) - c_{i_1 i_2 \cdots i_{q-1}y}\overline{D}_{i_q}(\xi_3).
\end{aligned}
\tag{31}
$$

Here

$$
\begin{aligned}
\overline{D}_\alpha = {} & \frac{\partial}{\partial \alpha} + a_{1,\alpha}\frac{\partial}{\partial a_1} + a_{2,\alpha}\frac{\partial}{\partial a_2} + b_{1,\alpha}\frac{\partial}{\partial b_1} + b_{2,\alpha}\frac{\partial}{\partial b_2} + c_\alpha\frac{\partial}{\partial c} + a_{1,\alpha\beta}\frac{\partial}{\partial a_{1,\beta}}\\
& + a_{2,\alpha\beta}\frac{\partial}{\partial a_{2,\beta}} + b_{1,\alpha\beta}\frac{\partial}{\partial b_{1,\beta}} + b_{2,\alpha\beta}\frac{\partial}{\partial b_{2,\beta}} + c_{\alpha\beta}\frac{\partial}{\partial c_\beta} + \cdots,
\end{aligned}
$$

$\alpha, \beta \in \{t, x, y\}$. (32)

3 Semi-invariants of Two and Three-Dimensional Parabolic Equations

Semi-invariants associated with Eq. (7) when the dependent variable undergoes a linear change of the type (20) can be find using an operator

$$
\mathbf{X_D} = -2a_1\eta_x\frac{\partial}{\partial b_1} - 2a_2\eta_y\frac{\partial}{\partial b_2} + (\eta_t - b_1\eta_x - b_2\eta_y - a_1\eta_{xx} - a_2\eta_{yy})\frac{\partial}{\partial c}, \tag{33}
$$

that characterizes the said infinitesimal change in coefficients of (7) and obtained by setting all $\xi_k = \xi_k(t, x, y, u)$ for $k = 1, 2, 3$, equal to zero in the infinitesimal coordinates of (14). By plugging (33) in the invariants criterion

$$
\mathbf{X}_D J(a_j, b_j, c) = 0, \quad j = 1, 2, \tag{34}
$$

we obtain

$$
-2a_1\eta_x\frac{\partial J}{\partial b_1} - 2a_2\eta_y\frac{\partial J}{\partial b_2} + (\eta_t - b_1\eta_x - b_2\eta_y - a_1\eta_{xx} - a_2\eta_{yy})\frac{\partial J}{\partial c} = 0. \tag{35}
$$

Further this equation yields a system of first order linear PDEs when coefficients of η_x, η_y, η_{xx}, and η_{yy} are equated to zero due to arbitrariness of η. We arrive at $\frac{\partial J}{\partial b_j} = 0$, and $\frac{\partial J}{\partial c} = 0$ that results in $J = J(a_j)$, as the zeroth order invariant of (7). Similarly, the first order invariants are obtainable from

$$
\mathbf{X}_D^{[1]} J(a_j, b_j, c, a_{j_t}, b_{j_t}, c_t, a_{j_x}, b_{j_x}, c_x, a_{j_y}, b_{j_y}, c_y) = 0, \tag{36}
$$

where $\mathbf{X}_D^{[1]}$ denotes the once extended generator (33) which reads as

$$\mathbf{X}_D^{[1]} = \mathbf{X}_D - \rho_1 \frac{\partial}{\partial b_{1_t}} - \rho_2 \frac{\partial}{\partial b_{2_t}} + \rho_3 \frac{\partial}{\partial c_t} - \rho_4 \frac{\partial}{\partial b_{1_x}} - \rho_5 \frac{\partial}{\partial b_{2_x}} + \rho_6 \frac{\partial}{\partial c_x} - \rho_7 \frac{\partial}{\partial b_{1_y}} - \rho_8 \frac{\partial}{\partial b_{2_y}} + \rho_9 \frac{\partial}{\partial c_y}, \tag{37}$$

where

$$\begin{aligned}
\rho_1 &= 2(a_1\eta_{tx} + a_{1_t}\eta_x),\\
\rho_2 &= 2(a_2\eta_{ty} + a_{2_t}\eta_y),\\
\rho_3 &= \eta_{tt} - b_1\eta_{tx} - b_2\eta_{ty} - a_1\eta_{txx} - a_2\eta_{tyy}\\
&\quad - a_{1_t}\eta_{xx} - a_{2_t}\eta_{yy} - b_{1_t}\eta_x - b_{2_t}\eta_y,\\
\rho_4 &= 2(a_1\eta_{xx} + a_{1_x}\eta_x),\\
\rho_5 &= 2(a_2\eta_{xy} + a_{2_x}\eta_y),\\
\rho_6 &= \eta_{tx} - b_1\eta_{xx} - b_2\eta_{xy} - a_1\eta_{xxx} - a_2\eta_{xyy}\\
&\quad - a_{1_x}\eta_{xx} - a_{2_x}\eta_{yy} - b_{1_x}\eta_x - b_{2_x}\eta_y,\\
\rho_7 &= 2(a_1\eta_{xy} + a_{1_y}\eta_x),\\
\rho_8 &= 2(a_2\eta_{yy} + a_{2_y}\eta_y),\\
\rho_9 &= \eta_{ty} - b_1\eta_{xy} - b_2\eta_{yy} - a_1\eta_{xxy} - a_2\eta_{yyy}\\
&\quad - a_{1_y}\eta_{xx} - a_{2_y}\eta_{yy} - b_{1_y}\eta_x - b_{2_y}\eta_y.
\end{aligned} \tag{38}$$

Equation (36) produces a system of linear PDEs on setting coefficients of all the first, second and third order derivatives of $\eta(t, x, y)$ to zero. Solving it we find a first order semi-invariant

$$\lambda_{11} := \frac{a_1^2(b_2a_{2_x} - a_2b_{2_x}) + a_2^2(a_1b_{1_y} - b_1a_{1_y})}{a_1a_2^2}, \tag{39}$$

and

$$J = J(a_j, a_{j_t}, a_{j_x}, a_{j_y}). \tag{40}$$

Where Eq. (40) implies that a_j and all its first order derivatives remain unaltered in both the equations which could be mapped into each other by means of an invertible point transformation (20) when (39) agrees for both of them. For derivation of semi-invariants of (7) under transformation of *only* the independent variables we use a subgroup (21) of the equivalence transformations. An operator

$$\begin{aligned}\mathbf{X_I} &= \xi_1(t)\frac{\partial}{\partial t} + \xi_2(t,x)\frac{\partial}{\partial x} + \xi_3(t,y)\frac{\partial}{\partial y} + a_1(2\xi_{2_x} - \xi_{1_t})\frac{\partial}{\partial a_1}\\ &+a_2(2\xi_{3_y} - \xi_{1_t})\frac{\partial}{\partial a_2} + (a_1\xi_{2_{xx}} + b_1(\xi_{2_x} - \xi_{1_t}) - \xi_{2_t})\frac{\partial}{\partial b_1}\\ &+(a_2\xi_{3_{yy}} + b_2(\xi_{3_y} - \xi_{1_t}) - \xi_{3_t})\frac{\partial}{\partial b_2} - c\xi_{1_t}\frac{\partial}{\partial c},\end{aligned} \tag{41}$$

characterizes the said infinitesimal change that is obtained from (14) by eliminating η, and its derivatives from the corresponding infinitesimal coordinates (19). The invariance criterion

$$\mathbf{X_I} J(a_j, b_j, c) = 0, \quad j = 1, 2, \tag{42}$$

leads to a trivial invariant, i.e., $J =$ constant. However, the first order invariants are as follows

$$\begin{aligned}&\lambda_{21} := \frac{a_2^{3/2} a_{1_y}}{a_1^{3/2} a_{2_x}}, \quad \lambda_{22} := \frac{a_2^2 c}{a_1 a_{2_x}^2}, \quad \lambda_{23} := \frac{a_2^3 c_x}{a_1 a_{2_x}^3}, \quad \lambda_{24} := \frac{a_2^{7/2} c_y}{a_1^{3/2} a_{2_x}^3},\\ &\lambda_{25} := \frac{\sqrt{a_2}(2a_2 b_{2_x} - a_{2_x} a_{2_y})}{\sqrt{a_1} a_{2_x}^2}, \quad \lambda_{26} := \frac{a_2^{5/2}(2a_1 b_{1_y} - a_{1_x} a_{1_y})}{a_1^{5/2} a_{2_x}^2},\end{aligned} \tag{43}$$

that are derived by applying first extension of (41), that is

$$\begin{aligned}\mathbf{X_I^{[1]}} &= \mathbf{X_I} + \rho_{10}\frac{\partial}{\partial a_{1_t}} + \rho_{11}\frac{\partial}{\partial a_{2_t}} + \rho_{12}\frac{\partial}{\partial b_{1_t}} + \rho_{13}\frac{\partial}{\partial b_{2_t}} - \rho_{14}\frac{\partial}{\partial c_t} + \rho_{15}\frac{\partial}{\partial a_{1_x}}\\ &+\rho_{16}\frac{\partial}{\partial a_{2_x}} + \rho_{17}\frac{\partial}{\partial b_{1_x}} + \rho_{18}\frac{\partial}{\partial b_{2_x}} - \rho_{19}\frac{\partial}{\partial c_x} + \rho_{20}\frac{\partial}{\partial a_{1_y}} + \rho_{21}\frac{\partial}{\partial a_{2_y}}\\ &+\rho_{22}\frac{\partial}{\partial b_{1_y}} + \rho_{23}\frac{\partial}{\partial b_{2_y}} - \rho_{24}\frac{\partial}{\partial c_y},\end{aligned} \tag{44}$$

with

$$\begin{aligned}\rho_{10} &= a_1(2\xi_{2_{tx}} - \xi_{1_{tt}}) + 2a_{1_t}(\xi_{2_x} - \xi_{1_t}) - a_{1_x}\xi_{2_t} - a_{1_y}\xi_{3_t},\\ \rho_{11} &= a_2(2\xi_{3_{ty}} - \xi_{1_{tt}}) + 2a_{2_t}(\xi_{3_y} - \xi_{1_t}) - a_{2_x}\xi_{2_t} - a_{2_y}\xi_{3_t},\\ \rho_{12} &= a_1\xi_{2_{txx}} + b_1(\xi_{2_{tx}} - \xi_{1_{tt}}) + a_{1_t}\xi_{2_{xx}} + b_{1_t}(\xi_{2_x} - 2\xi_{1_t})\\ &\quad - b_{1_x}\xi_{2_t} - b_{1_y}\xi_{3_t} - \xi_{2_{tt}},\\ \rho_{13} &= a_2\xi_{3_{tyy}} + b_2(\xi_{3_{ty}} - \xi_{1_{tt}}) + a_{2_t}\xi_{3_{yy}} + b_{2_t}(\xi_{3_y} - 2\xi_{1_t})\\ &\quad - b_{2_x}\xi_{2_t} - b_{2_y}\xi_{3_t} - \xi_{3_{tt}},\\ \rho_{14} &= c\xi_{1_{tt}} + 2c_t\xi_{1_t} + c_x\xi_{2_t} + c_y\xi_{3_t},\\ \rho_{15} &= 2a_1\xi_{2_{xx}} + a_{1_x}(\xi_{2_x} - \xi_{1_t}),\\ \rho_{16} &= a_{2_x}(2\xi_{3_y} - \xi_{1_t} - \xi_{2_x}),\end{aligned}$$

$$\rho_{17} = a_1\xi_{2_{xxx}} + b_1\xi_{2_{xx}} + a_{1_x}\xi_{2_{xx}} - b_{1_x}\xi_{1_t} - \xi_{2_{tx}},$$
$$\rho_{18} = a_{2_x}\xi_{3_{yy}} + b_{2_x}(\xi_{3_y} - \xi_{1_t} - \xi_{2_x}),$$
$$\rho_{19} = c_x(\xi_{1_t} + \xi_{2_x}),$$
$$\rho_{20} = a_{1_y}(2\xi_{2_x} - \xi_{1_t} - \xi_{3_y}),$$
$$\rho_{21} = 2a_2\xi_{3_{yy}} + a_{2_y}(\xi_{3_y} - \xi_{1_t}),$$
$$\rho_{22} = a_{1_y}\xi_{2_{xx}} + b_{1_y}(\xi_{2_x} - \xi_{1_t} - \xi_{3_y}),$$
$$\rho_{23} = a_2\xi_{3_{yyy}} + b_2\xi_{3_{yy}} + a_{2_y}\xi_{3_{yy}} - b_{2_y}\xi_{1_t} - \xi_{3_{ty}},$$
$$\rho_{24} = c_y(\xi_{1_t} + \xi_{3_y}), \tag{45}$$

on $J(a_j, b_j, c, a_{j_t}, b_{j_t}, c_t, a_{j_x}, b_{j_x}, c_x, a_{j_y}, b_{j_y}, c_y)$.

Likewise, semi-invariants of the parabolic equation with three spatial dimensions (8) are derived. Firstly we look for semi-invariants associated with (8) due to a linear infinitesimal change of dependent variable u. Generator (22) enables deduction of the said invariants after restricting its infinitesimal coordinates (24) to only contain η and its derivatives, which reads as

$$\mathbf{Z_D} = -2a_1\eta_x\frac{\partial}{\partial b_1} - 2a_2\eta_y\frac{\partial}{\partial b_2} - 2a_3\eta_z\frac{\partial}{\partial b_2} + (\eta_t - b_1\eta_x - b_2\eta_y - b_3\eta_z$$
$$-a_1\eta_{xx} - a_2\eta_{yy} - a_3\eta_{zz})\frac{\partial}{\partial c}. \tag{46}$$

Considering the invariance test

$$\mathbf{Z_D}J(a_k, b_k, c) = 0, \quad k = 1, 2, 3, \tag{47}$$

and solving the obtained system we find nonexistence of zero order invariant for (8) except

$$J = J(a_k), \tag{48}$$

that implies a_k, for $k = 1, 2, 3$, should remain the same for both the equations that can be transformed into each other by means of point transformations. Extending (46) once we get

$$\mathbf{Z_D^{[1]}} = \mathbf{Z_D} - \rho_1\frac{\partial}{\partial b_{1_t}} - \rho_2\frac{\partial}{\partial b_{2_t}} - \rho_3\frac{\partial}{\partial b_{3_t}} + \rho_4\frac{\partial}{\partial c_t} - \rho_5\frac{\partial}{\partial b_{1_x}} - \rho_6\frac{\partial}{\partial b_{2_x}}$$
$$-\rho_7\frac{\partial}{\partial b_{3_x}} + \rho_8\frac{\partial}{\partial c_x} - \rho_9\frac{\partial}{\partial b_{1_y}} - \rho_{10}\frac{\partial}{\partial b_{2_y}} - \rho_{11}\frac{\partial}{\partial b_{3_y}} + \rho_{12}\frac{\partial}{\partial c_y}$$
$$-\rho_{13}\frac{\partial}{\partial b_{1_z}} - \rho_{14}\frac{\partial}{\partial b_{2_z}} - \rho_{15}\frac{\partial}{\partial b_{3_z}} + \rho_{16}\frac{\partial}{\partial c_z}, \tag{49}$$

with

$$\begin{aligned}
\rho_1 &= 2(a_1\eta_{tx} + a_{1_t}\eta_x),\\
\rho_2 &= 2(a_2\eta_{ty} + a_{2_t}\eta_y),\\
\rho_3 &= 2(a_3\eta_{tz} + a_{3_t}\eta_z),\\
\rho_4 &= \eta_{tt} - a_1\eta_{txx} - a_2\eta_{tyy} - a_3\eta_{tzz} - b_1\eta_{tx} - b_2\eta_{ty} - b_3\eta_{tz}\\
&\quad - a_{1_t}\eta_{xx} - a_{2_t}\eta_{yy} - a_{3_t}\eta_{zz} - b_{1_t}\eta_x - b_{2_t}\eta_y - b_{3_t}\eta_z,\\
\rho_5 &= 2(a_1\eta_{xx} + a_{1_x}\eta_x),\\
\rho_6 &= 2(a_2\eta_{xy} + a_{2_x}\eta_y),\\
\rho_7 &= 2(a_3\eta_{xz} + a_{3_x}\eta_z),\\
\rho_8 &= \eta_{tx} - a_1\eta_{xxx} - a_2\eta_{xyy} - a_3\eta_{xzz} - b_1\eta_{xx} - b_2\eta_{xy} - b_3\eta_{xz}\\
&\quad - a_{1_x}\eta_{xx} - a_{2_x}\eta_{yy} - a_{3_x}\eta_{zz} - b_{1_x}\eta_x - b_{2_x}\eta_y - b_{3_x}\eta_z,\\
\rho_9 &= 2(a_1\eta_{xy} + a_{1_y}\eta_x),\\
\rho_{10} &= 2(a_2\eta_{yy} + a_{2_y}\eta_y),\\
\rho_{11} &= 2(a_3\eta_{yz} + a_{3_y}\eta_z),\\
\rho_{12} &= \eta_{ty} - a_1\eta_{xxy} - a_2\eta_{yyy} - a_3\eta_{yzz} - b_1\eta_{xy} - b_2\eta_{yy} - b_3\eta_{yz}\\
&\quad - a_{1_y}\eta_{xx} - a_{2_y}\eta_{yy} - a_{3_y}\eta_{zz} - b_{1_y}\eta_x - b_{2_y}\eta_y - b_{3_y}\eta_z,\\
\rho_{23} &= 2(a_1\eta_{xz} + a_{1_z}\eta_x),\\
\rho_{14} &= 2(a_2\eta_{yz} + a_{2_z}\eta_y),\\
\rho_{15} &= 2(a_3\eta_{zz} + a_{3_z}\eta_z),\\
\rho_{16} &= \eta_{tz} - a_1\eta_{xxz} - a_2\eta_{yyz} - a_3\eta_{zzz} - b_1\eta_{xz} - b_2\eta_{yz} - b_3\eta_{zz}\\
&\quad - a_{1_z}\eta_{xx} - a_{2_z}\eta_{yy} - a_{3_z}\eta_{zz} - b_{1_z}\eta_x - b_{2_z}\eta_y - b_{3_z}\eta_z,
\end{aligned} \tag{50}$$

which enables derivation of the associated first order invariants. We apply $\mathbf{Z_D^{[1]}}$, i.e.,

$$\mathbf{Z_D^{[1]}} J(a_k, b_k, c, a_{k_t}, b_{k_t}, c_t, a_{k_x}, b_{k_x}, c_x, a_{k_y}, b_{k_z}, c_y, a_{k_z}, b_{k_z}, c_z) = 0, \tag{51}$$

which splits into a system of linear PDEs when all the coefficients of η_{xxx}, η_{xyy}, η_{xxy}, η_{xzz}, η_{yyy}, η_{yzz}, η_{tt}, η_{tx}, η_{ty}, η_{tz}, η_{xx}, η_{xy}, η_{xz}, η_{yy}, η_{yz}, η_{zz}, η_t, η_x, η_y and η_z, are equated to zero. Solution of obtained system provides the following semi-invariants

$$\begin{aligned}
\lambda_{31} &:= \frac{a_1^2(b_2a_{2_x} - a_2b_{2_x}) + a_2^2(a_1b_{1_y} - b_1a_{1_y})}{a_1a_2^2},\\
\lambda_{32} &:= \frac{a_1^2(b_3a_{3_x} - a_3b_{3_x}) + a_3^2(a_1b_{1_z} - b_1a_{1_z})}{a_1a_3^2},\\
\lambda_{33} &:= \frac{a_2^2(b_3a_{3_y} - a_3b_{3_y}) + a_3^2(a_2b_{2_z} - b_2a_{2_z})}{a_2a_3^2},
\end{aligned} \tag{52}$$

with

$$J = J(a_k, a_{k_t}, a_{k_x}, a_{k_y}, a_{k_z}, c, c_x, c_y), \tag{53}$$

as invariant quantities.

Semi-invariants associated with (8) due to a change of the independent variables are derivable from

$$\begin{aligned}
\mathbf{Z_I} J = \; & \xi_1(t)\frac{\partial J}{\partial t} + \xi_2(t,x)\frac{\partial J}{\partial x} + \xi_3(t,y)\frac{\partial J}{\partial y} + \xi_4(t,z)\frac{\partial J}{\partial z} + \eta(t,x,y,z)u\frac{\partial J}{\partial u} \\
& + a_1(2\xi_{2_x} - \xi_{1_t})\frac{\partial J}{\partial a_1} + a_2(2\xi_{3_y} - \xi_{1_t})\frac{\partial J}{\partial a_2} + a_3(2\xi_{4_z} - \xi_{1_t})\frac{\partial J}{\partial a_3} \\
& + (a_1(\xi_{2_{xx}} - 2\eta_x) + b_1(\xi_{2_x} - \xi_{1_t}) - \xi_{2_t})\frac{\partial J}{\partial b_1} + (a_2(\xi_{3_{yy}} - 2\eta_y) \\
& + b_2(\xi_{3_y} - \xi_{1_t}) - \xi_{3_t})\frac{\partial J}{\partial b_2} + (a_3(\xi_{4_{zz}} - 2\eta_z) + b_3(\xi_{4_z} - \xi_{1_t}) - \xi_{4_t})\frac{\partial J}{\partial b_2} \\
& + (\eta_t - a_1\eta_{xx} - a_2\eta_{yy} - a_3\eta_{zz} - b_1\eta_x - b_2\eta_y - b_3\eta_z - c\xi_{1_t})\frac{\partial J}{\partial c} = 0,
\end{aligned} \tag{54}$$

where $\mathbf{Z_I}$, is obtained by eliminating η and its derivatives from its infinitesimal coordinates (24). On equating coefficients of all derivatives of ξ_l, for $l = 1, 2, 3, 4$, in (54) to zero leads us to a system of linear PDEs that solves to generate J =constant. Furthermore, a system of linear PDEs

$$\begin{aligned}
& \frac{\partial J}{\partial a_{1_t}} = 0, \; \frac{\partial J}{\partial a_{2_t}} = 0, \; \frac{\partial J}{\partial a_{3_t}} = 0, \; \frac{\partial J}{\partial b_1} = 0, \; \frac{\partial J}{\partial b_2} = 0, \; \frac{\partial J}{\partial b_3} = 0, \; \frac{\partial J}{\partial b_{1_t}} = 0, \\
& \frac{\partial J}{\partial b_{2_t}} = 0, \; \frac{\partial J}{\partial b_{3_t}} = 0, \; \frac{\partial J}{\partial b_{1_x}} = 0, \; \frac{\partial J}{\partial b_{2_y}} = 0, \; \frac{\partial J}{\partial b_{3_z}} = 0, \; \frac{\partial J}{\partial c_t} = 0, \\
& 2a_1\frac{\partial J}{\partial a_{1_x}} + a_{1_y}\frac{\partial J}{\partial b_{1_y}} + a_{1_z}\frac{\partial J}{\partial b_{1_z}} = 0, \\
& 2a_2\frac{\partial J}{\partial a_{2_y}} + a_{2_x}\frac{\partial J}{\partial b_{2_x}} + a_{2_z}\frac{\partial J}{\partial b_{2_z}} = 0, \\
& 2a_3\frac{\partial J}{\partial a_{3_z}} + a_{3_x}\frac{\partial J}{\partial b_{3_x}} + a_{3_y}\frac{\partial J}{\partial b_{3_y}} = 0, \\
& 2a_1\frac{\partial J}{\partial a_1} + a_{1_x}\frac{\partial J}{\partial a_{1_x}} - a_{2_x}\frac{\partial J}{\partial a_{2_x}} - a_{3_x}\frac{\partial J}{\partial a_{3_x}} + 2a_{1_y}\frac{\partial J}{\partial a_{1_y}} + 2a_{1_z}\frac{\partial J}{\partial a_{1_z}} \\
& -b_{2_x}\frac{\partial J}{\partial b_{2_x}} - b_{3_x}\frac{\partial J}{\partial b_{3_x}} + b_{1_y}\frac{\partial J}{\partial b_{1_y}} + b_{1_z}\frac{\partial J}{\partial b_{1_z}} - c_x\frac{\partial J}{\partial c_x} = 0, \\
& 2a_2\frac{\partial J}{\partial a_2} - a_{1_y}\frac{\partial J}{\partial a_{1_y}} + 2a_{2_x}\frac{\partial J}{\partial a_{2_x}} + a_{2_y}\frac{\partial J}{\partial a_{2_y}} + 2a_{2_z}\frac{\partial J}{\partial a_{2_z}} - a_{3_y}\frac{\partial J}{\partial a_{3_y}} \\
& -b_{1_y}\frac{\partial J}{\partial b_{1_y}} + b_{2_x}\frac{\partial J}{\partial b_{2_x}} + b_{2_z}\frac{\partial J}{\partial b_{2_z}} - b_{3_y}\frac{\partial J}{\partial b_{3_y}} - c_y\frac{\partial J}{\partial c_y} = 0, \\
& 2a_3\frac{\partial J}{\partial a_3} - a_{1_z}\frac{\partial J}{\partial a_{1_z}} - a_{2_z}\frac{\partial J}{\partial a_{2_z}} + 2a_{3_x}\frac{\partial J}{\partial a_{3_x}} + 2a_{3_y}\frac{\partial J}{\partial a_{3_y}} + a_{3_z}\frac{\partial J}{\partial a_{3_z}}
\end{aligned}$$

$$-b_{1_z}\frac{\partial J}{\partial b_{1_z}} - b_{2_z}\frac{\partial J}{\partial b_{2_z}} + b_{3_x}\frac{\partial J}{\partial b_{3_x}} + b_{3_y}\frac{\partial J}{\partial b_{3_y}} - c_z\frac{\partial J}{\partial c_z} = 0,$$

$$a_1\frac{\partial J}{\partial a_1} + a_2\frac{\partial J}{\partial a_2} + a_3\frac{\partial J}{\partial a_3} + c\frac{\partial J}{\partial c} + a_{1_x}\frac{\partial J}{\partial a_{1_x}} + a_{1_y}\frac{\partial J}{\partial a_{1_y}} + a_{1_z}\frac{\partial J}{\partial a_{1_z}}$$
$$+a_{2_x}\frac{\partial J}{\partial a_{2_x}} + a_{2_y}\frac{\partial J}{\partial a_{2_y}} + a_{2_z}\frac{\partial J}{\partial a_{2_z}} + a_{3_x}\frac{\partial J}{\partial a_{3_x}} + a_{3_y}\frac{\partial J}{\partial a_{3_y}} + a_{3_z}\frac{\partial J}{\partial a_{3_z}}$$
$$+b_{1_y}\frac{\partial J}{\partial b_{1_y}} + b_{1_z}\frac{\partial J}{\partial b_{1_z}} + b_{2_x}\frac{\partial J}{\partial b_{2_x}} + b_{2_z}\frac{\partial J}{\partial b_{2_z}} + b_{3_x}\frac{\partial J}{\partial b_{3_x}} + b_{3_y}\frac{\partial J}{\partial b_{3_y}}$$
$$+c_x\frac{\partial J}{\partial c_x} + c_y\frac{\partial J}{\partial c_y} + c_z\frac{\partial J}{\partial c_z} = 0, \tag{55}$$

surfaces from

$$\mathbf{Z}_\mathbf{I}^{[1]} J(a_k, b_k, c, a_{k_t}, b_{k_t}, c_t, a_{k_x}, b_{k_x}, c_x, a_{k_y}, b_{k_z}, c_y, a_{k_z}, b_{k_z}, c_z) = 0, \tag{56}$$

when all the coefficients of ξ_l and its derivatives of all orders are equated to zero. Solution of the system (55) reveals following first order semi-invariants

$$\lambda_{41} := \frac{a_2 a_{3_x}}{a_3 a_{2_x}}, \quad \lambda_{42} := \frac{a_2^{3/2} a_{1_y}}{a_1^{3/2} a_{2_x}}, \quad \lambda_{43} := \frac{a_2^{3/2} a_{3_y}}{\sqrt{a_1} a_3 a_{2_x}}, \quad \lambda_{44} := \frac{a_2\sqrt{a_3} a_{1_z}}{a_1^{3/2} a_{2_x}},$$
$$\lambda_{45} := \frac{\sqrt{a_3} a_{2_z}}{\sqrt{a_1} a_{2_x}}, \quad \lambda_{46} := \frac{\sqrt{a_2}(2a_2 b_{2_x} - a_{2_x} a_{2_y})}{\sqrt{a_1} a_{2_x}^2}, \quad \lambda_{47} := \frac{a_2^2(2a_3 b_{3_x} - a_{3_x} a_{3_z})}{\sqrt{a_1} a_3^{3/2} a_{2_x}^2},$$
$$\lambda_{48} := \frac{a_2^{5/2}(2a_1 b_{1_y} - a_{1_x} a_{1_y})}{a_1^{5/2} a_{2_x}^2}, \quad \lambda_{49} := \frac{a_2^{5/2}(2a_3 b_{3_y} - a_{3_y} a_{3_z})}{a_1 a_3^{3/2} a_{2_x}^2},$$
$$\lambda_{50} := \frac{a_2^2\sqrt{a_3}(2a_1 b_{1_z} - a_{1_x} a_{1_z})}{a_1^{5/2} a_{2_x}^2}, \quad \lambda_{51} := \frac{\sqrt{a_2}\sqrt{a_3}(2a_2 b_{2_z} - a_{2_y} a_{2_z})}{a_1 a_{2_x}^2}$$
$$\lambda_{52} := \frac{a_2^2 c}{a_1 a_{2_x}^2}, \quad \lambda_{53} := \frac{a_2^3 c_x}{a_1 a_{2_x}^3}, \quad \lambda_{54} := \frac{a_2^{7/2} c_y}{a_1^{3/2} a_{2_x}^3}, \quad \lambda_{55} := \frac{a_2^3\sqrt{a_3} c_z}{a_1^{3/2} a_{2_x}^3}. \tag{57}$$

4 Invertible Point Transformations for Families of Two and Three-Dimensional Parabolic Equations

Equivalence transformations for two and three-dimensional parabolic PDEs have been derived in Sect. 2. An algorithm is presented here to obtain the invertible point transformations to map two and three-dimensional parabolic equations into equations of the same dimensions when corresponding semi-invariants agree. Considering first Eq. (7) and transforming the dependent variable linearly (20). Under this transformation (7) gets into the same form but with new coefficients $\overline{a}_j, \overline{b}_j, \overline{c}$ that are given below

$$
\begin{aligned}
\overline{a}_j &= a_j, \\
\overline{b}_1 &- b_1 + 2a_1 \frac{\eta_x}{\eta}, \\
\overline{b}_2 &= b_2 + 2a_2 \frac{\eta_y}{\eta}, \\
\overline{c} &= c + \frac{a_1\eta_{xx} + a_2\eta_{yy} + b_1\eta_x + b_2\eta_y - \eta_t}{\eta}.
\end{aligned} \tag{58}
$$

Similarly, the most general transformations of the independent variables that map the parabolic two-dimensional equation to an equation of the same type are given in (21). Incorporating it in (7) leads us to an equation with the coefficients $\overline{a}_j, \overline{b}_j, \overline{c}$ given by

$$
\begin{aligned}
\overline{a}_1 &= a_1 \frac{\xi_{1_t}}{\xi_{2_x}^2}, \\
\overline{a}_2 &= a_2 \frac{\xi_{1_t}}{\xi_{3_y}^2}, \\
\overline{b}_1 &= \frac{(b_1\xi_{1_t} + \xi_{2_t})\xi_{2_x}^2 - a_1\xi_{2_{xx}}\xi_{1_t}}{\xi_{2_x}^3}, \\
\overline{b}_2 &= \frac{(b_2\xi_{1_t} + \xi_{3_t})\xi_{3_y}^2 - a_2\xi_{3_{yy}}\xi_{1_t}}{\xi_{3_y}^3}, \\
\overline{c} &= c\xi_{1_t}.
\end{aligned} \tag{59}
$$

The relationships (58) and (59) provides the algorithms to work out invertible transformations to map two-dimensional parabolic equations into each other for which the derived semi-invariants remain unaltered.

For the three-dimensional case the coefficients of the parabolic equation (8), under linear transformation of the dependent variable

$$
\overline{u} = \psi(t, x, y, z, u) := \eta(t, x, y, z)u, \tag{60}
$$

and the following change of independent variables

$$
\begin{aligned}
\overline{t} &= \phi_1(t, x, y, u) := \xi_1(t), \\
\overline{x} &= \phi_2(t, x, y, u) := \xi_2(t, x), \\
\overline{y} &= \phi_3(t, x, y, u) := \xi_3(t, y), \\
\overline{z} &= \phi_4(t, x, y, u) := \xi_4(t, z),
\end{aligned} \tag{61}
$$

become $\overline{a}_k, \overline{b}_k, \overline{c}$ given by

$$\begin{aligned}
\overline{a}_k &= a_k,\\
\overline{b}_1 &= b_1 + 2a_1\frac{\eta_x}{\eta},\\
\overline{b}_2 &= b_2 + 2a_2\frac{\eta_y}{\eta},\\
\overline{b}_3 &= b_3 + 2a_3\frac{\eta_z}{\eta},\\
\overline{c} &= c + \frac{a_1\eta_{xx} + a_2\eta_{yy} + a_3\eta_{zz} + b_1\eta_x + b_2\eta_y + b_3\eta_z - \eta_t}{\eta},
\end{aligned} \tag{62}$$

and

$$\begin{aligned}
\overline{a}_1 &= a_1\frac{\xi_{1_t}}{\xi_{2_x}^2},\\
\overline{a}_2 &= a_2\frac{\xi_{1_t}}{\xi_{3_y}^2},\\
\overline{a}_3 &= a_3\frac{\xi_{1_t}}{\xi_{4_z}^2},\\
\overline{b}_1 &= \frac{(b_1\xi_{1_t} + \xi_{2_t})\xi_{2_x}^2 - a_1\xi_{2_{xx}}\xi_{1_t}}{\xi_{2_x}^3},\\
\overline{b}_2 &= \frac{(b_2\xi_{1_t} + \xi_{3_t})\xi_{3_y}^2 - a_2\xi_{3_{yy}}\xi_{1_t}}{\xi_{3_y}^3},\\
\overline{b}_3 &= \frac{(b_3\xi_{1_t} + \xi_{4_t})\xi_{4_z}^2 - a_3\xi_{4_{zz}}\xi_{1_t}}{\xi_{4_z}^3},\\
\overline{c} &= c\xi_{1_t},
\end{aligned} \tag{63}$$

respectively. Now we present application of the obtained semi-invariants.

5 Applications

1. The following parabolic two-dimensional equations

$$u_t - u_{xx} - u_{yy} - \frac{4}{x}u_x - \frac{2}{y}u_y - \frac{2}{x^2}u = 0, \tag{64}$$

and

$$\overline{u}_t - \overline{u}_{xx} - \overline{u}_{yy} = 0, \tag{65}$$

are transformable into each other as is ensured by the semi-invariants (39), i.e., for both of them $\lambda_{11} = 0$. Existence of invertible point transformation to map given equations into each other is therefore guaranteed, for which we employ (58) to get

$$u = \frac{1}{x^2 y}\overline{u}. \tag{66}$$

2. A parabolic equation of the form

$$u_t - yu_{xx} - xu_{yy} + 2yu_x - \frac{2x}{y}u_y - \left(\frac{1+2ty}{t}\right)u = 0, \tag{67}$$

is transformable to a relatively simple linear two-dimensional parabolic equation

$$\overline{u}_t - y\overline{u}_{xx} - x\overline{u}_{yy} - y\overline{u} = 0, \tag{68}$$

via an invertible point transformation

$$u = \frac{e^x}{ty}\overline{u}. \tag{69}$$

Coefficients of both the equations when inserted in (39) gives the same result.

3. Consider the following linear partial differential equation in four independent variables

$$u_t + c_1(u_x + u_{xx}) + c_2(u_y + u_{yy}) - \frac{4\alpha_t + (c_1 + c_2)\alpha(t)}{\alpha(t)}(u_z + u_{zz}) + \frac{2\alpha_t + \alpha(t)}{\alpha(t)}u = 0. \tag{70}$$

It has $\lambda_{31} = \lambda_{32} = \lambda_{33} = 0$ and could be mapped to a parabolic three-dimensional equation

$$\overline{u}_t + c_1\overline{u}_{xx} + c_2\overline{u}_{yy} - \frac{4\alpha_t + (c_1 + c_2)\alpha(t)}{\alpha(t)}\overline{u}_{zz} + \frac{2\alpha_t + \alpha(t)}{\alpha(t)}\overline{u} = 0, \tag{71}$$

since all invariants (52) are zero. Invertible point transformation which relates both the equations is

$$u = \alpha(t)e^{-\frac{1}{2}(x+y+z)}\overline{u}. \tag{72}$$

4. An equation of the form

$$u_t + \frac{1}{t^2}u_{xx} + \frac{x}{4ty^2}u_{yy} - \frac{x}{t}u_x - \frac{1}{2}\left(\frac{y}{t} + \frac{x}{2ty^3}\right)u_y + u = 0, \tag{73}$$

transforms to

$$u_{\bar{t}} + u_{\bar{x}\bar{x}} + \bar{x}u_{\bar{y}\bar{y}} + u = 0, \tag{74}$$

via

$$t = \bar{t}, \ \ x = \frac{\bar{x}}{\bar{t}}, \ \ y = \sqrt{\frac{\bar{y}}{\bar{t}}}. \tag{75}$$

These transformations along with the coefficients of both the PDEs (73) and (74) generate $\lambda_{21} = \frac{1}{x}$, $\lambda_{22} = \lambda_{23} = \lambda_{24} = \lambda_{25} = \lambda_{26} = 0$.

5. The linear parabolic equation with four independent variables

$$u_t - \frac{x^2}{t^2}u_{xx} + t^3 \ln x u_{yy} + z^2 u_{zz} - \left(\frac{x}{t^2} + \frac{x \ln x}{t}\right) u_x + \frac{y}{t}u_y + zu_z + u = 0, \tag{76}$$

is reducible to the following simplest linear form

$$u_{\bar{t}} - u_{\bar{x}\bar{x}} + \bar{x}u_{\bar{y}\bar{y}} + u_{\bar{z}\bar{z}} + u = 0, \tag{77}$$

under the transformations

$$t = \bar{t}, \ \ x = \exp\left(\frac{\bar{x}}{\bar{t}}\right), \ \ y = \bar{t}\bar{y}, \ \ z = \exp(\bar{z}). \tag{78}$$

All the invariants (57) agree for both the equations under (78).

6 Conclusion

Lie Infinitesimal method has been employed to derive semi-invariants of the two and three-dimensional parabolic PDEs. Equivalence transformations for both the cases are obtained for which we employed MAPLE, as for linear PDEs and systems of such equations the said package has a built in code that constructs the group of equivalence transformations. These equivalence transformations lead to first order semi-invariants of the two and three dimensional parabolic equations associated due to invertible transformations of the dependent and independent variables separately. Deduced invariants ensure existence of invertible point transformations between two equations if they agree for both of them, this fact enabled to get mappings which transform the corresponding linear parabolic equations of dimension two and three to much simpler forms, in applications.

References

1. Ibragimov, N.H.: Equivalence Groups and Invariants of Linear and Nonlinear Equations, 1st edn. Arch ALGA (2004)
2. Olver, P.J.: Invariants and Symmetry. Cambridge University Press, Cambridge (1995)
3. Ovsiannikov, L.V.: Group Analysis of Differential Equations. Academic Press, New York (1982)
4. Laplace, P.S.: Laplaces Cuevres Completes, vol. IX, p. 5. Gauthier- Villars, Paris (1893). English translation, New York, 1966
5. Cotton, E.: Sur les invariants differentiels dde quelques equations lineaires aux dérivées partiellelles du second ordre, vol. 17, p. 211. Annales Scientifique de l'École Normale Supérieure (1900)
6. Ibragimov, N.H.: Nonlinear Dyn. **28**
7. Johnpillai, I.K., Mahomed, F.M.: J. Phys. A: Math. Gen. **34**
8. Johnpillai, I.K., Mahomed, F.M., Wafo Soh, C.: J. Nonlinear. Math. Phys. **9**
9. Mahomed, F.M.: J. Nonlinear. Math. Phys. **15**
10. Mahomed, F.M., Pooe, C.A.: Proceedings of the International Conference Modern Group Analysis, vol. 9. Moscow State University, Moscow
11. Aslam, A., Mahomed, F.M.: Sci. World J. **2013**
12. Mahomed, F.M., Qadir, A., Ramnarain, A.: Math. Prob. Eng. **2011**
13. Mahomed, F.M., Safdar, M., Zama, J.: Commun. Nonlinear. Sci. Numer. Simul. **17**
14. Tsaousi, C., Sophocleous, C.: J. Math. Anal. Appl. **363**
15. Tsaousi, C., Sophocleous, C., Tracina, R.: J. Math. Anal. Appl. **349**
16. Lie, S.: Arch. Mat. Naturvidenskab **9**, Reprinted in Lie's Ges. Abhandl **5**
17. Ibragimov, N.H.: Elementary Lie Group Analysis and Ordinary Differential Equations. Wiley, New York (1999)

Kaleidoscope of Classical Vortex Images and Quantum Coherent States

Oktay K. Pashaev and Aygül Koçak

Abstract The Schrödinger cat states, constructed from Glauber coherent states and applied for description of qubits are generalized to the kaleidoscope of coherent states, related with regular n-polygon symmetry and the roots of unity. This quantum kaleidoscope is motivated by our method of classical hydrodynamics images in a wedge domain, described by q-calculus of analytic functions with q as a primitive root of unity. First we treat in detail the trinity states and the quartet states as descriptive for qutrit and ququat units of quantum information. Normalization formula for these states requires introduction of specific combinations of exponential functions with mod 3 and mod 4 symmetry, which are known also as generalized hyperbolic functions. We show that these states can be generated for an arbitrary n by the Quantum Fourier transform and can provide in general, qudit unit of quantum information. Relations of our states with quantum groups and quantum calculus are discussed.

Keywords Coherent states · Quantum information · Qubit · Qutrit
Qudit · Quantum Fourier transform

1 Introduction

1.1 Classical Vortex Kaleidoscope

The classical problem of point vortices in a domain bounded by two infinite circular cylinders with arbitrary radiuses and positions in the plane, can be formulated as the Apollonius circles problem, reducible by Möbius transformation to the one in annular domain between two concentric circles [1]. Recently we have formulated the two circles theorem, allowing one to construct an arbitrary flow in such annular domain

O. K. Pashaev · A. Koçak (✉)
Department of Mathematics, Izmir Institute of Technology, 35430 Izmir, Turkey
e-mail: oktaypashaev@iyte.edu.tr

A. Koçak
e-mail: aygulkocak@iyte.edu.tr

V. G. Kac et al. (eds.), *Symmetries, Differential Equations and Applications*,
Springer Proceedings in Mathematics & Statistics 266,
https://doi.org/10.1007/978-3-030-01376-9_10

by the complex potential $F(z)$ as q-periodic analytic function, $F(qz) = F(z)$, where $q = R^2/r^2$ is determined by ratio of two circle radiuses. Depending on the number and the position of vortices, sources or sinks, one can fix singularities of this function in terms of q-elementary functions [2]. Similar theorem [3] formulated for the flow in the wedge domain with angle $\frac{\pi}{n}$, requires construction of complex potential $F(z)$ as q^2-periodic function, $F(q^2z) = F(z)$ with q as a root of unity $q^{2n} = 1$. It determines complex velocity $\overline{V}(z) = dF(z)/dz$ as q^2- self-similar analytic function

$$\overline{V}(q^2z) = q^{-2}\,\overline{V}(z).$$

The wedge theorem describes the fluid flow as superposition of complex analytic functions

$$F(z) = \sum_{k=0}^{n-1} f(q^{2k}z) + \sum_{k=0}^{n-1} \bar{f}(q^{2k}z)\,, \tag{1}$$

representing the kaleidoscope of images associated with the regular $2n$ - polygon. For the point vortex located at z_0, the theorem gives q^2-periodic complex potential

$$F(z) = \frac{i\Gamma}{2\pi} \ln \frac{z^n - z_0^n}{z^n - \bar{z}_0^n} = F(q^2z), \tag{2}$$

which due to the Kummer expansion

$$z^n - z_0^n = (z - z_0)(z - q^2z_0)(z - q^4z_0)\ldots(z - q^{2(n-1)}z_0)\,,$$

appears as the set of vortices with even images at points $z_0,\ q^2z_0,\ q^4z_0,\ \ldots,\ q^{2(n-1)}z_0$ and with odd images at $\bar{z}_0,\ q^2\bar{z}_0,\ q^4\bar{z}_0,\ \ldots,\ q^{2(n-1)}\bar{z}_0$. This kaleidoscope of vortex images we called the Kummer kaleidoscope.

1.2 Quantum Kaleidoscope and Coherent States

Since analytical functions are related intrinsically with quantum coherent states and the Fock–Bargman representation, here we extend our ideas to the Hilbert space for the coherent states. The problem is to construct q-periodic quantum states and q-self-similar quantum states. Similar problem, relating self-similarity properties of fractals, the theory of entire analytical functions and the q-deformed algebra with coherent states was discussed recently in [4]. In the present paper we consider the case, when q is the primitive root of unity $q^{2n} = 1$ and show that it leads to the kaleidoscope of coherent states $|\alpha\rangle$, $|q^2\alpha\rangle$, ..., $|q^{2(n-1)}\alpha\rangle$, located at vertices of the regular polygon. By acting with dilatation operator on analytic function

$$f(q^2z) = q^{2z\frac{d}{dz}} f(z) \tag{3}$$

we can rewrite the wedge theorem (1) in a compact form

$$F(z) = \sum_{k=0}^{n-1} \left(q^{2z\frac{d}{dz}}\right)^k [f(z) + \bar{f}(z)] = [n]_{q^{2z\frac{d}{dz}}} [f(z) + \bar{f}(z)], \tag{4}$$

where we have used non-symmetric $\hat{Q}$-number

$$[n]_{\hat{Q}} = 1 + \hat{Q} + \hat{Q}^2 + \cdots \hat{Q}^{n-1} = \frac{\hat{Q}^n - 1}{\hat{Q} - 1},$$

with the operator base

$$\hat{Q} \equiv q^{2z\frac{d}{dz}}. \tag{5}$$

From this representation, q^2-periodicity of function $F(z)$ follows easily. Due to the identity $\hat{Q}^n = q^{2nz\frac{d}{dz}} = 1$ we have

$$\hat{Q}[n]_{\hat{Q}} = [n]_{\hat{Q}}$$

and as follows

$$F(q^2 z) = \hat{Q}F(z) = \hat{Q}[n]_{\hat{Q}}[f(z) + \bar{f}(z)] = [n]_{\hat{Q}}[f(z) + \bar{f}(z)] = F(z).$$

It is noticed that the differential operator (5) is the Fock–Bargman representation for the dilatation operator $\hat{Q} = q^{2\widehat{N}}$, acting on coherent states as

$$q^{2\widehat{N}}|\alpha\rangle = |q^2\alpha\rangle, \tag{6}$$

where $\widehat{N} = \hat{a}^\dagger \hat{a}$ is the number operator. Then, by analogy with the wedge theorem (1) and (4), we can construct q^2-periodic quantum state as superposition of coherent states

$$\begin{aligned} |0\rangle_\alpha &\equiv |\alpha\rangle + |q^2\alpha\rangle + |q^4\alpha\rangle + \cdots + |q^{2(n-1)}\alpha\rangle \\ &= (I + q^{2\widehat{N}} + q^{4\widehat{N}} + \cdots + q^{2(n-1)\widehat{N}})|\alpha\rangle = [n]_{q^{2\widehat{N}}}|\alpha\rangle. \end{aligned} \tag{7}$$

The q^2-periodicity for this quantum state

$$q^{2\widehat{N}}|0\rangle_\alpha = |0\rangle_\alpha$$

follows easily from the relation $\hat{Q}^n = q^{2n\widehat{N}} = I$. This suggests also that to find q^2-self-similar quantum states, one can take the following superpositions of coherent states

$$|1\rangle_\alpha \equiv [n]_{q^{2\widehat{N}+2}}|\alpha\rangle, \quad |2\rangle_\alpha \equiv [n]_{q^{2\widehat{N}+4}}|\alpha\rangle, \quad \ldots \quad , \quad |n-1\rangle_\alpha \equiv [n]_{q^{2\widehat{N}+2(n-1)}}|\alpha\rangle,$$

satisfying self-similarity conditions

$$q^{2\widehat{N}}|1\rangle_\alpha = q^2|1\rangle_\alpha, \;\; q^{2\widehat{N}}|2\rangle_\alpha = q^4|2\rangle_\alpha, \ldots, \;\; q^{2\widehat{N}}|n-1\rangle_\alpha = q^{2(n-1)}|n-1\rangle_\alpha.$$

It turns out that this construction provides the set of orthogonal quantum states. The similar superpositions of coherent states were discussed in different context by several authors, as the generalized coherent states [5, 6], as factorization problem for the Schrödinger equation with self-similar potential [7] and as the Schrödinger cat states [8]. The Schrödinger cat states [9] as superposition of Glauber's optical coherent states with opposite phases, become important tool for construction of qubits, as a units of quantum information [10] in quantum optics [11]. They correspond to even and odd quantum states with $q^2 = -1$. Here we generalize this construction to the kaleidoscope of coherent states, related with regular n-polygon symmetry and the roots of unity. Superposition of coherent states with such symmetry plays the role of the quantum Fourier transform and provides the set of orthonormal quantum states, as a description of qutrits, ququats and qudits. Such quantum states, considered as a units of quantum information processing and corresponding to an arbitrary base number n, could have advantage in secure quantum communication.

1.3 Glauber Coherent States

We consider the Heisenberg–Weyl algebra, written in terms of creation and annihilation operators, satisfying bosonic commutation relation

$$\left[\hat{a}, \hat{a}^\dagger\right] = \widehat{\mathrm{I}}.$$

The annihilation operator determines the vacuum state $\hat{a}|0\rangle = 0$ from the Hilbert space $|0\rangle \in H$ and the creation operator $\hat{a}^\dagger$ repeatedly applied to this state, gives orthonormal set of states $|n\rangle = \dfrac{(\hat{a}^\dagger)^n}{\sqrt{n!}}|0\rangle$. Coherent states are defined as eigenstates of annihilation operator [12]:

$$\hat{a}|\alpha\rangle = \alpha|\alpha\rangle\,,$$

where $\alpha \in C$. This gives us a relation between complex plane and the Hilbert space, such that $\alpha \in C \leftrightarrow |\alpha\rangle \in H$. Another equivalent definition is given by the displacement operator,

$$D(\alpha) = e^{\alpha\hat{a}^\dagger - \bar{\alpha}\hat{a}} = e^{-\frac{1}{2}|\alpha|^2} e^{\alpha\hat{a}^\dagger} e^{-\bar{\alpha}\hat{a}} \tag{8}$$

so that,

$$|\alpha\rangle = D(\alpha)|0\rangle = e^{-\frac{1}{2}|\alpha|^2} \sum_{n=0}^{\infty} \frac{\alpha^n}{\sqrt{n!}}|n\rangle\,. \tag{9}$$

From this we get the following representation of coherent states:

$$|\alpha\rangle = \frac{e^{\alpha \hat{a}^\dagger}}{\sqrt{e^{|\alpha|^2}}}|0\rangle,$$

which is instructive for our generalizations. The inner product of coherent states,

$$\langle\alpha|\beta\rangle = e^{-\frac{1}{2}|\alpha|^2-\frac{1}{2}|\beta|^2+\bar{\alpha}\beta}$$

is never zero, $|\langle\alpha|\beta\rangle|^2 = e^{-|\alpha-\beta|^2} \neq 0$. This is why coherent states are not orthogonal. The aim of the present paper is to construct an orthogonal set of states as superposition of coherent states with discrete regular polygon symmetry.

2 Schrödinger's Cat States

In description of the Schrödinger cat states one introduces two orthogonal states as superpositions of $|\alpha\rangle$ and $|-\alpha\rangle$ states, which are called even and odd cat states [8],

$$|\mathrm{Cat}_{even}\rangle \sim |\alpha\rangle + |-\alpha\rangle \quad , \quad |\mathrm{Cat}_{odd}\rangle \sim |\alpha\rangle - |-\alpha\rangle .$$

The states in this superpositions are related by rotation to angle π, which corresponds to primitive root of unity $q^2 = \bar{q}^2 = -1$, so that $q^4 = 1$. The normalization constants for these states

$$|0\rangle_\alpha = \frac{N_0}{\sqrt{2}}(\,|\alpha\rangle + |q^2\alpha\rangle\,), \quad |1\rangle_\alpha = \frac{N_1}{\sqrt{2}}(\,|\alpha\rangle + \bar{q}^2|q^2\alpha\rangle\,), \tag{10}$$

are calculated as:

$$N_0 = \frac{e^{\frac{|\alpha|^2}{2}}}{\sqrt{2}\sqrt{\cosh|\alpha|^2}}, \quad N_1 = \frac{e^{\frac{|\alpha|^2}{2}}}{\sqrt{2}\sqrt{\sinh|\alpha|^2}}. \tag{11}$$

Transformation to these states can be described in the matrix form as an action by the Hadamard gate,

$$\begin{bmatrix} |0\rangle_\alpha \\ |1\rangle_\alpha \end{bmatrix} = \mathbf{N} \underbrace{\frac{1}{\sqrt{2}}\begin{bmatrix} 1 & 1 \\ 1 & \bar{q}^2 \end{bmatrix}}_{Hadamard\ gate} \begin{bmatrix} |\alpha\rangle \\ |q^2\alpha\rangle \end{bmatrix}, \tag{12}$$

where the normalization matrix

$$\mathbf{N} = \frac{e^{\frac{|\alpha|^2}{2}}}{\sqrt{2}} \operatorname{diag}\left({}_0e^{|\alpha|^2},\ {}_1e^{|\alpha|^2}\right)^{-1/2} (mod\, 2) \equiv \operatorname{diag}(N_0, N_1) \tag{13}$$

is defined by the even (0 mod 2) and the odd (1 mod 2) exponential functions, coinciding with hyperbolic functions,

$$(mod\, 2)\ \ {}_0e^{|\alpha|^2} \equiv \sum_{k=0}^{\infty} \frac{\left(|\alpha|^2\right)^{2k}}{(2k)!} = \frac{e^{|\alpha|^2} + e^{q^2|\alpha|^2}}{2} = \cosh|\alpha|^2,$$

$$(mod\, 2)\ \ {}_1e^{|\alpha|^2} \equiv \sum_{k=0}^{\infty} \frac{\left(|\alpha|^2\right)^{2k+1}}{(2k+1)!} = \frac{e^{|\alpha|^2} + \bar{q}^2 e^{q^2|\alpha|^2}}{2} = \sinh|\alpha|^2.$$

2.1 Mod 2 Representation of Cat States

In terms of these exponential functions we can rewrite the Schrödinger cat states in a compact form:

$$|0\rangle_\alpha = \frac{{}_0e^{\alpha\hat{a}^\dagger}}{\sqrt{{}_0e^{|\alpha|^2}}}|0\rangle\ \ (mod\, 2) = \frac{\cosh\alpha\hat{a}^\dagger}{\sqrt{\cosh|\alpha|^2}}|0\rangle\,,$$

$$|1\rangle_\alpha = \frac{{}_1e^{\alpha\hat{a}^\dagger}}{\sqrt{{}_1e^{|\alpha|^2}}}|0\rangle\ \ (mod\, 2) = \frac{\sinh\alpha\hat{a}^\dagger}{\sqrt{\sinh|\alpha|^2}}|0\rangle.$$

2.2 Eigenvalue Problem for Cat States

Since $|\alpha\rangle$ is an eigenstate of annihilation operator $\hat{a}$, $\hat{a}|\alpha\rangle = \alpha|\alpha\rangle$, it is also the eigenstate of operator $\hat{a}^2$:

$$\hat{a}^2|\alpha\rangle = \alpha^2|\alpha\rangle.$$

However, the last equation admits one more eigenstate $|-\alpha\rangle$ with the same eigenvalue α^2, so that

$$\hat{a}^2|\mp\alpha\rangle = \alpha^2|\mp\alpha\rangle.$$

Hence, any superposition of states $\{|+\alpha\rangle, |-\alpha\rangle\}$ is also an eigenstate of operator $\hat{a}^2$, with the same eigenvalue. This implies that Schrödinger cat states are eigenstates of this operator,

$$\hat{a}^2|0\rangle_\alpha = \alpha^2|0\rangle_\alpha,\quad \hat{a}^2|1\rangle_\alpha = \alpha^2|1\rangle_\alpha,$$

constituting orthonormal basis $\{|0\rangle_\alpha, |1\rangle_\alpha\}$. It can be used to define the qubit coherent state:

$$|\psi\rangle_\alpha = c_0|0\rangle_\alpha + c_1|1\rangle_\alpha,$$

where $|c_0|^2 + |c_1|^2 = 1$, representing a unit of quantum information in quantum optics. This qubit state is an eigenstate of operator $\hat{a}^2$ as well:

$$\hat{a}^2|\psi\rangle_\alpha = \alpha^2|\psi\rangle_\alpha.$$

2.3 Number of Photons in Cat States

The cat states are not eigenstates of the annihilation operator $\hat{a}$. On the contrary, action of this operator gives flipping between cat states $|0\rangle_\alpha$ and $|1\rangle_\alpha$:

$$\hat{a}|0\rangle_\alpha = \alpha\frac{N_0}{N_1}|1\rangle_\alpha, \quad \hat{a}|1\rangle_\alpha = \alpha\frac{N_1}{N_0}|0\rangle_\alpha.$$

By using these equations we find number of photons in Schrödinger's cat states as :

$$\begin{aligned}
{}_\alpha\langle 0|\widehat{N}|0\rangle_\alpha &= |\alpha|^2\frac{N_0^2}{N_1^2} = |\alpha|^2\frac{{}_1e^{|\alpha|^2}}{{}_0e^{|\alpha|^2}} = |\alpha|^2\tanh|\alpha|^2,\\
{}_\alpha\langle 1|\widehat{N}|1\rangle_\alpha &= |\alpha|^2\frac{N_1^2}{N_0^2} = |\alpha|^2\frac{{}_0e^{|\alpha|^2}}{{}_1e^{|\alpha|^2}} = |\alpha|^2\coth|\alpha|^2.
\end{aligned}$$

It shows deviation from number of photons in coherent states

$$\langle\alpha|\widehat{N}|\alpha\rangle = |\alpha|^2$$

shown in Fig. 1. In the limiting case $|\alpha| \to \infty$ both distributions asymptotically goes to this value

$$\lim_{|\alpha|\to\infty} {}_\alpha\langle 0|\widehat{N}|0\rangle_\alpha = \lim_{|\alpha|\to\infty} {}_\alpha\langle 1|\widehat{N}|1\rangle_\alpha \approx |\alpha|^2.$$

The cat states for $|\alpha|^2 \ll 1$ are reduced to the so called Schrödinger's kitten states with number of photons 0 and 1:

$$\lim_{|\alpha|\to 0} {}_\alpha\langle 0|\widehat{N}|0\rangle_\alpha = 0, \quad \lim_{|\alpha|\to 0} {}_\alpha\langle 1|\widehat{N}|1\rangle_\alpha = 1.$$

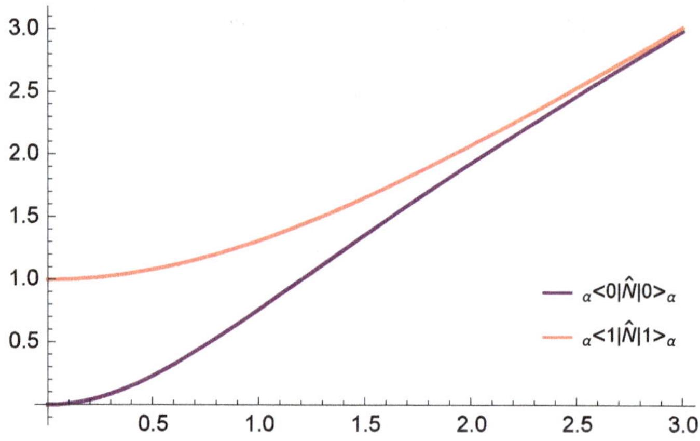

Fig. 1 Photon numbers in Schrödinger's cat states

2.4 *Fermionic Representation of Cat States*

The dilatation operator $q^{2\widehat{N}} = e^{i\pi\widehat{N}} = (-1)^{\widehat{N}}$ is the parity operator for cat states, so that $|0\rangle_\alpha$ and $|1\rangle_\alpha$ states are eigenstates of this operator. The first state is the q^2-periodic state and the second one is q^2-self-similar state,

$$q^{2\widehat{N}}|0\rangle_\alpha = |0\rangle_\alpha\,, \quad q^{2\widehat{N}}|1\rangle_\alpha = q^2|1\rangle_\alpha. \tag{14}$$

These states represent kaleidoscope of two coherent states $|\alpha\rangle$ and $|-\alpha\rangle$, rotated by angle π, and can be rewritten in terms of parity operator

$$\begin{aligned}|0\rangle_\alpha &= N_0\,[2]_{q^{2\widehat{N}}}|\alpha\rangle = N_0(I + q^{2\widehat{N}})|\alpha\rangle,\\ |1\rangle_\alpha &= N_1\,[2]_{q^{2\widehat{N}+2}}|\alpha\rangle = N_1(I + q^2 q^{2\widehat{N}})|\alpha\rangle,\end{aligned} \tag{15}$$

or

$$\begin{aligned}|0\rangle_\alpha &= N_0(I + (-1)^{\widehat{N}})|\alpha\rangle\,,\\ |1\rangle_\alpha &= N_1(I - (-1)^{\widehat{N}})|\alpha\rangle\,.\end{aligned} \tag{16}$$

It is noticed that the cat states are eigenstates also of q^2- non-symmetric number operator

$$[\widehat{N}]_{q^2} = \frac{q^{2\widehat{N}} - 1}{q^2 - 1},$$

where $q^2 = -1$,

$$[\widehat{N}]_{q^2}|0\rangle_\alpha = [0]_{q^2}|0\rangle_\alpha, \qquad [\widehat{N}]_{q^2}|1\rangle_\alpha = [1]_{q^2}|1\rangle_\alpha,$$

with eigenvalues $[0]_{q^2} = 0$ and $[1]_{q^2} = 1$. In the Fock basis $|n\rangle$, $n = 0, 1, 2...$, these number operator is diagonal, with eigenvalues 0 for even numbers $n = 2k$, and 1 for odd numbers $n = 2k + 1$. This number operator in the cat basis is matrix of the fermion number operator

$$[\widehat{N}]_{q^2} = \begin{pmatrix} 0 & 0 \\ 0 & 1 \end{pmatrix} = \widehat{N}_F$$

factorized by fermionic creation and annihilation operators $\widehat{N}_F = \hat{b}^\dagger \hat{b}$, with algebra

$$\hat{b}\hat{b}^\dagger + \hat{b}^\dagger \hat{b} = I, \qquad \hat{b}^2 = 0, \qquad (\hat{b}^\dagger)^2 = 0,$$

and matrix representation

$$\hat{b} = \begin{pmatrix} 0 & 1 \\ 0 & 0 \end{pmatrix}, \qquad \hat{b}^\dagger = \begin{pmatrix} 0 & 0 \\ 1 & 0 \end{pmatrix}.$$

The cat states in this basis then are just computational basis qubit states:

$$|0\rangle = \begin{pmatrix} 1 \\ 0 \end{pmatrix}, \qquad |1\rangle = \begin{pmatrix} 0 \\ 1 \end{pmatrix}.$$

3 Trinity States

The Schrödinger cat states can be generalized to the kaleidoscope of coherent states. We start this generalization from the set of three coherent states, rotated by angle $\frac{2\pi}{3}$ and located at vertices of equilateral triangle, which corresponds to roots of unity $q^6 = 1$. First we define superposition

$$|0\rangle_\alpha = \frac{N_0}{\sqrt{3}} \left(|\alpha\rangle + |q^2\alpha\rangle + |q^4\alpha\rangle \right).$$

Due to identity

$$q^{6n} - 1 = (q^{2n} - 1)(1 + q^{2n} + q^{4n}) = 0 \Rightarrow$$

$$1 + q^{2n} + q^{4n} = 3\,\delta_{n,0(mod\,3)},$$

with

$$\delta_{k,0\,(mod\,3)} = \begin{cases} 1, \; k = 0\,(mod\,3); \\ 0, \; k \neq 0\,(mod\,3), \end{cases} \tag{17}$$

the normalization constant is $N_0 = e^{\frac{|\alpha|^2}{2}}(3\,{}_0e^{|\alpha|^2})^{-1/2}$, where we have introduced $(mod\ 3)$ exponential function

$$ {}_0e^{|\alpha|^2}(mod\ 3) \equiv \sum_{k=0}^{\infty} \frac{\left(|\alpha|^2\right)^{3k}}{(3k)!} = \frac{1}{3}\left(e^{|\alpha|^2} + e^{q^2|\alpha|^2} + e^{q^4|\alpha|^2}\right). $$

In a similar way we obtain the set of orthonormal states $|0\rangle_\alpha$, $|1\rangle_\alpha$ and $|2\rangle_\alpha$:

$$ |0\rangle_\alpha = e^{\frac{|\alpha|^2}{2}} \frac{|\alpha\rangle + |q^2\alpha\rangle + |q^4\alpha\rangle}{\sqrt{3}\sqrt{e^{|\alpha|^2} + e^{q^2|\alpha|^2} + e^{q^4|\alpha|^2}}} = e^{\frac{|\alpha|^2}{2}} \frac{|\alpha\rangle + |q^2\alpha\rangle + |q^4\alpha\rangle}{3\sqrt{{}_0e^{|\alpha|^2}(mod\ 3)}}, $$

$$ |1\rangle_\alpha = e^{\frac{|\alpha|^2}{2}} \frac{|\alpha\rangle + \bar{q}^2|q^2\alpha\rangle + \bar{q}^4|q^4\alpha\rangle}{\sqrt{3}\sqrt{e^{|\alpha|^2} + \bar{q}^2 e^{q^2|\alpha|^2} + \bar{q}^4 e^{q^4|\alpha|^2}}} = e^{\frac{|\alpha|^2}{2}} \frac{|\alpha\rangle + \bar{q}^2|q^2\alpha\rangle + \bar{q}^4|q^4\alpha\rangle}{3\sqrt{{}_1e^{|\alpha|^2}(mod\ 3)}}, $$

$$ |2\rangle_\alpha = e^{\frac{|\alpha|^2}{2}} \frac{|\alpha\rangle + \bar{q}^4|q^2\alpha\rangle + \bar{q}^2|q^4\alpha\rangle}{\sqrt{3}\sqrt{e^{|\alpha|^2} + \bar{q}^4 e^{q^2|\alpha|^2} + \bar{q}^2 e^{q^4|\alpha|^2}}} = e^{\frac{|\alpha|^2}{2}} \frac{|\alpha\rangle + \bar{q}^4|q^2\alpha\rangle + \bar{q}^2|q^4\alpha\rangle}{3\sqrt{{}_2e^{|\alpha|^2}(mod\ 3)}}. $$

3.1 Matrix Form of Trinity States

These states appear by action of the trinity gate, playing the role of three dimensional analogue of Hadamard gate

$$ \begin{bmatrix} |0\rangle_\alpha \\ |1\rangle_\alpha \\ |2\rangle_\alpha \end{bmatrix} = \underbrace{\mathbf{N}\frac{1}{\sqrt{3}}\begin{bmatrix} 1 & 1 & 1 \\ 1 & \bar{q}^2 & (\bar{q}^2)^2 \\ 1 & \bar{q}^4 & (\bar{q}^4)^2 \end{bmatrix}}_{Trinity\ gate} \begin{bmatrix} |\alpha\rangle \\ |q^2\alpha\rangle \\ |q^4\alpha\rangle \end{bmatrix}, \tag{18} $$

with normalization constants

$$ \mathbf{N} = \frac{e^{\frac{|\alpha|^2}{2}}}{\sqrt{3}} \operatorname{diag}\left({}_0e^{|\alpha|^2}, {}_1e^{|\alpha|^2}, {}_2e^{|\alpha|^2}\right)^{-1/2} (mod\ 3) \equiv \operatorname{diag}(N_0, N_1, N_2) \tag{19} $$

and identity

$$ 1 + \bar{q}^{2(n-k)} + \bar{q}^{4(n-k)} = 3\,\delta_{n,k(mod\ 3)},\ 0 \le k \le 2, $$

where

$$ \delta_{n,k\,(mod\ 3)} = \begin{cases} 1,\ n = k\ (mod\ 3); \\ 0,\ n \ne k\ (mod\ 3). \end{cases} \tag{20} $$

Trinity states as superposition of coherent states have the following explicit phase shift :

$$|0\rangle_\alpha = N_0(|\alpha\rangle + |e^{i\frac{2\pi}{3}}\alpha\rangle + |e^{-i\frac{2\pi}{3}}\alpha\rangle),$$
$$|1\rangle_\alpha = N_1(|\alpha\rangle + e^{-i\frac{2\pi}{3}}|e^{i\frac{2\pi}{3}}\alpha\rangle + e^{i\frac{2\pi}{3}}|e^{-i\frac{2\pi}{3}}\alpha\rangle),$$
$$|2\rangle_\alpha = N_2(|\alpha\rangle + e^{i\frac{2\pi}{3}}|e^{i\frac{2\pi}{3}}\alpha\rangle + e^{-i\frac{2\pi}{3}}|e^{-i\frac{2\pi}{3}}\alpha\rangle).$$

By using three different (*mod* 3) exponential functions, we can rewrite these states in a compact form:

$$|0\rangle_\alpha = \frac{{}_0e^{\alpha\hat{a}^\dagger}}{\sqrt{{}_0e^{|\alpha|^2}}}|0\rangle\,,\quad |1\rangle_\alpha = \frac{{}_1e^{\alpha\hat{a}^\dagger}}{\sqrt{{}_1e^{|\alpha|^2}}}|0\rangle\,,\quad |2\rangle_\alpha = \frac{{}_2e^{\alpha\hat{a}^\dagger}}{\sqrt{{}_2e^{|\alpha|^2}}}|0\rangle \qquad (mod\ 3).$$

3.2 Eigenvalue Problem for Trinity States

Coherent states $\{|\alpha\rangle, |q^2\alpha\rangle, |q^4\alpha\rangle\}$ are eigenstates of operator $\hat{a}$ with different eigenvalues $\alpha, q^2\alpha, q^4\alpha$, and the eigenstates of operator $\hat{a}^3$ with the same eigenvalue α^3. Due to this, our trinity states $\{|0\rangle_\alpha, |1\rangle_\alpha, |2\rangle_\alpha\}$ are also eigenstates of operator $\hat{a}^3$:

$$\hat{a}^3|q^{2k}\alpha\rangle = \alpha^3|q^{2k}\alpha\rangle \quad \Rightarrow \quad \hat{a}^3|k\rangle_\alpha = \alpha^3|k\rangle_\alpha,\quad k = 0, 1, 2.$$

From trinity states we can construct the qutrit coherent state

$$|\psi\rangle_\alpha = c_0|0\rangle_\alpha + c_1|1\rangle_\alpha + c_2|2\rangle_\alpha,$$

where $|c_0|^2 + |c_1|^2 + |c_2|^2 = 1$, as a unit of quantum information with base 3. It turns out that this state is an eigenstate of operator $\hat{a}^3$:

$$\hat{a}^3|\psi\rangle_\alpha = \alpha^3|\psi\rangle_\alpha.$$

3.3 Number of Photons in Trinity States

The annihilation operator $\hat{a}$ acts on states $|0\rangle_\alpha$, $|1\rangle_\alpha$ and $|2\rangle_\alpha$ as cyclic permutation:

$$\hat{a}|0\rangle_\alpha = \alpha\frac{N_0}{N_2}|2\rangle_\alpha\,,\ \hat{a}|1\rangle_\alpha = \alpha\frac{N_1}{N_0}|0\rangle_\alpha\,,\ \hat{a}|2\rangle_\alpha = \alpha\frac{N_2}{N_1}|1\rangle_\alpha\,. \tag{21}$$

This equation allows us to calculate number of photons in trinity states (see Fig. 2):

$${}_\alpha\langle 0|\widehat{N}|0\rangle_\alpha = |\alpha|^2\left[\frac{{}_2e^{|\alpha|^2}}{{}_0e^{|\alpha|^2}}\right] = |\alpha|^2\left[\frac{1 + 2e^{\frac{-3|\alpha|^2}{2}}\cos\left(\frac{\sqrt{3}}{2}|\alpha|^2 + \frac{2\pi}{3}\right)}{1 + 2e^{\frac{-3|\alpha|^2}{2}}\cos\left(\frac{\sqrt{3}}{2}|\alpha|^2\right)}\right],$$

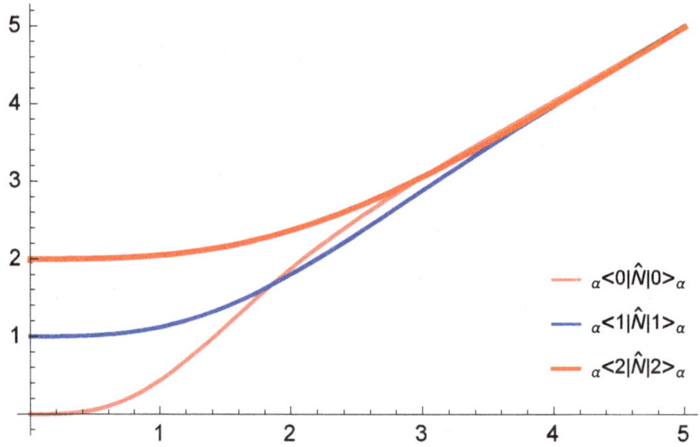

Fig. 2 Photon numbers in trinity states

$$ {}_{\alpha}\langle 1|\widehat{N}|1\rangle_{\alpha} = |\alpha|^2 \left[\frac{{}_0e^{|\alpha|^2}}{{}_1e^{|\alpha|^2}}\right] = |\alpha|^2 \left[\frac{1+2e^{\frac{-3|\alpha|^2}{2}}\cos\left(\frac{\sqrt{3}}{2}|\alpha|^2\right)}{1+2e^{\frac{-3|\alpha|^2}{2}}\cos\left(\frac{\sqrt{3}}{2}|\alpha|^2-\frac{2\pi}{3}\right)}\right], $$

$$ {}_{\alpha}\langle 2|\widehat{N}|2\rangle_{\alpha} = |\alpha|^2 \left[\frac{{}_1e^{|\alpha|^2}}{{}_2e^{|\alpha|^2}}\right] = |\alpha|^2 \left[\frac{1+2e^{\frac{-3|\alpha|^2}{2}}\cos\left(\frac{\sqrt{3}}{2}|\alpha|^2-\frac{2\pi}{3}\right)}{1+2e^{\frac{-3|\alpha|^2}{2}}\cos\left(\frac{\sqrt{3}}{2}|\alpha|^2+\frac{2\pi}{3}\right)}\right]. $$

3.3.1 Matrix Representation

Due to $\widehat{N}|n\rangle = n|n\rangle, n \geq 0$ from the eigenvalue problem

$$ q^{2\widehat{N}}|0\rangle_{\alpha} = |0\rangle_{\alpha}, \quad q^{2\widehat{N}}|1\rangle_{\alpha} = q^2|1\rangle_{\alpha}, \quad q^{2\widehat{N}}|2\rangle_{\alpha} = q^4|2\rangle_{\alpha}\,, $$

we find the matrix representation of operators in our kaleidoscope basis as the clock and the shift matrix

$$ q^{2\widehat{N}} = \begin{pmatrix} 1 & 0 & 0 \\ 0 & q^2 & 0 \\ 0 & 0 & q^4 \end{pmatrix}, \quad \hat{a} = \alpha \begin{pmatrix} 0 & \frac{N_1}{N_0} & 0 \\ 0 & 0 & \frac{N_2}{N_1} \\ \frac{N_0}{N_2} & 0 & 0 \end{pmatrix} = \alpha \begin{pmatrix} \frac{N_1}{N_0} & 0 & 0 \\ 0 & \frac{N_2}{N_1} & 0 \\ 0 & 0 & \frac{N_0}{N_2} \end{pmatrix} \begin{pmatrix} 0 & 1 & 0 \\ 0 & 0 & 1 \\ 1 & 0 & 0 \end{pmatrix}. \quad (22) $$

This gives for the q^2-number operator $[\widehat{N}]_{q^2} = \frac{q^{2\widehat{N}}-1}{q^2-1}$, the diagonal form with matrix elements

$$_{\alpha}\langle 0|[\widehat{N}]_{q^2}|0\rangle_{\alpha} = [0]_{q^2}, \quad _{\alpha}\langle 1|[\widehat{N}]_{q^2}|1\rangle_{\alpha} = [1]_{q^2}, \quad _{\alpha}\langle 2|[\widehat{N}]_{q^2}|2\rangle_{\alpha} = [2]_{q^2},$$

as q^2 numbers: $[0]_{q^2} = 0$, $[1]_{q^2} = 1$, $[2]_{q^2} = \frac{1+i\sqrt{3}}{2}$.

4 Quartet States

We define four states, rotated by angle $\frac{\pi}{2}$ and determined by primitive roots of unity: $q^8 = 1$. Superposition of these states with proper coefficients give us quartet of orthonormal basis states:

$$\begin{bmatrix} |0\rangle_\alpha \\ |1\rangle_\alpha \\ |2\rangle_\alpha \\ |3\rangle_\alpha \end{bmatrix} = \mathbf{N} \underbrace{\frac{1}{\sqrt{4}} \begin{bmatrix} 1 & 1 & 1 & 1 \\ 1 & \bar{q}^2 & (\bar{q}^2)^2 & (\bar{q}^2)^3 \\ 1 & \bar{q}^4 & (\bar{q}^4)^2 & (\bar{q}^4)^3 \\ 1 & \bar{q}^6 & (\bar{q}^6)^2 & (\bar{q}^6)^3 \end{bmatrix}}_{Quartet\ gate} \begin{bmatrix} |\alpha\rangle \\ |q^2\alpha\rangle \\ |q^4\alpha\rangle \\ |q^6\alpha\rangle \end{bmatrix}, \tag{23}$$

where normalization constants are defined as

$$\mathbf{N} = \frac{e^{\frac{|\alpha|^2}{2}}}{\sqrt{4}} \operatorname{diag}\left({}_0e^{|\alpha|^2}, {}_1e^{|\alpha|^2}, {}_2e^{|\alpha|^2}, {}_3e^{|\alpha|^2}\right)^{-1/2} (mod\, 4) \equiv \operatorname{diag}(N_0, N_1, N_2, N_3)$$

and the identity is

$$1 + \bar{q}^{2(n-k)} + \bar{q}^{4(n-k)} + \bar{q}^{6(n-k)} = 4\,\delta_{n,k(mod\, 4)} \quad , 0 \le k \le 3$$

with

$$\delta_{n,k\,(mod\, 4)} = \begin{cases} 1, & n = k\ (mod\, 4); \\ 0, & n \neq k\ (mod\, 4). \end{cases} \tag{24}$$

The quartet states are superpositions of cat states with explicit form of phase shift as

$$\begin{aligned} |0\rangle_\alpha &= N_0\left[(|\alpha\rangle + |-\alpha\rangle) + (|i\alpha\rangle + |-i\alpha\rangle)\right], \\ |1\rangle_\alpha &= N_1\left[(|\alpha\rangle - |-\alpha\rangle) - i(|i\alpha\rangle - |-i\alpha\rangle)\right], \\ |2\rangle_\alpha &= N_2\left[(|\alpha\rangle + |-\alpha\rangle) - (|i\alpha\rangle + |-i\alpha\rangle)\right], \\ |3\rangle_\alpha &= N_3\left[(|\alpha\rangle - |-\alpha\rangle) + i(|i\alpha\rangle - |-i\alpha\rangle)\right]. \end{aligned}$$

By using $(mod\, 4)$ exponential functions we get representation of these states in a compact form:

$$|0\rangle_\alpha = \frac{{}_0e^{\alpha\hat{a}^\dagger}}{\sqrt{{}_0e^{|\alpha|^2}}}|0\rangle, \quad |1\rangle_\alpha = \frac{{}_1e^{\alpha\hat{a}^\dagger}}{\sqrt{{}_1e^{|\alpha|^2}}}|0\rangle, \quad |2\rangle_\alpha = \frac{{}_2e^{\alpha\hat{a}^\dagger}}{\sqrt{{}_2e^{|\alpha|^2}}}|0\rangle, \quad |3\rangle_\alpha = \frac{{}_3e^{\alpha\hat{a}^\dagger}}{\sqrt{{}_3e^{|\alpha|^2}}}|0\rangle .$$

4.1 Eigenvalue Problem for Quartet States

As easy to see, the quartet states are eigenstates of operator $\hat{a}^4$ with eigenvalue α^4:

$$\hat{a}^4|q^{2k}\alpha\rangle = \alpha^4|q^{2k}\alpha\rangle \quad \Rightarrow \quad \hat{a}^4|k\rangle_\alpha = \alpha^4|k\rangle_\alpha \quad k = 0, 1, 2, 3.$$

As a result, the ququat state, defined as

$$|\psi\rangle_\alpha = c_0|0\rangle_\alpha + c_1|1\rangle_\alpha + c_2|2\rangle_\alpha + c_3|3\rangle_\alpha,$$

where $|c_0|^2 + |c_1|^2 + |c_2|^2 + |c_3|^2 = 1$, describes a unit of quantum information with base 4, and is an eigenstate of operator $\hat{a}^4$:

$$\hat{a}^4|\psi\rangle_\alpha = \alpha^4|\psi\rangle_\alpha.$$

4.2 Number of Photons in Quartet States

The annihilation operator $\hat{a}$ implements cyclic permutation of states $|k\rangle_\alpha$, $k = 0, 1, 2, 3$:

$$\hat{a}|0\rangle_\alpha = \alpha\frac{N_0}{N_3}|3\rangle_\alpha\,, \quad \hat{a}|1\rangle_\alpha = \alpha\frac{N_1}{N_0}|0\rangle_\alpha\,, \quad \hat{a}|2\rangle_\alpha = \alpha\frac{N_2}{N_1}|1\rangle_\alpha\,, \quad \hat{a}|3\rangle_\alpha = \alpha\frac{N_3}{N_2}|2\rangle_\alpha\,,$$

allowing us to calculate number of photons in quartet states (See Fig. 3):

$$
\begin{aligned}
{}_\alpha\langle 0|\widehat{N}|0\rangle_\alpha &= |\alpha|^2\left[\frac{{}_3e^{|\alpha|^2}}{{}_0e^{|\alpha|^2}}\right] = |\alpha|^2\left[\frac{\sinh|\alpha|^2 - \sin|\alpha|^2}{\cosh|\alpha|^2 + \cos|\alpha|^2}\right],\\
{}_\alpha\langle 1|\widehat{N}|1\rangle_\alpha &= |\alpha|^2\left[\frac{{}_0e^{|\alpha|^2}}{{}_1e^{|\alpha|^2}}\right] = |\alpha|^2\left[\frac{\cosh|\alpha|^2 + \cos|\alpha|^2}{\sinh|\alpha|^2 + \sin|\alpha|^2}\right],\\
{}_\alpha\langle 2|\widehat{N}|2\rangle_\alpha &= |\alpha|^2\left[\frac{{}_1e^{|\alpha|^2}}{{}_2e^{|\alpha|^2}}\right] = |\alpha|^2\left[\frac{\sinh|\alpha|^2 + \sin|\alpha|^2}{\cosh|\alpha|^2 - \cos|\alpha|^2}\right],\\
{}_\alpha\langle 3|\widehat{N}|3\rangle_\alpha &= |\alpha|^2\left[\frac{{}_2e^{|\alpha|^2}}{{}_3e^{|\alpha|^2}}\right] = |\alpha|^2\left[\frac{\cosh|\alpha|^2 - \cos|\alpha|^2}{\sinh|\alpha|^2 - \sin|\alpha|^2}\right].
\end{aligned}
$$

The quartet states are also eigenstates of q^2-number operator $[\widehat{N}]_{q^2}$:

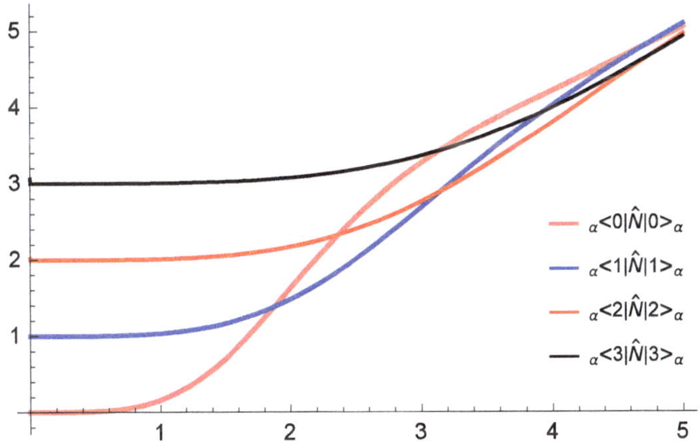

Fig. 3 Photon numbers in quartet states

$$q^{2\widehat{N}}|k\rangle_\alpha = q^{2k}|k\rangle_\alpha \quad \Rightarrow \quad [\widehat{N}]_{q^2}|k\rangle_\alpha = [k]_{q^2}|k\rangle_\alpha \text{ , where } k = 0, 1, 2, 3. \tag{25}$$

5 Kaleidoscope of Quantum Coherent States

As a generalization of previous results, here we consider superposition of n coherent states, which are belonging to vertices of regular n-polygon, rotated by angle $\frac{\pi}{n}$ (Fig. 4). It is related with primitive roots of unity: $q^{2n} = 1$. For the inner product of q^{2k} rotated coherent states we have

$$\langle q^{2k}\alpha|q^{2k}\alpha\rangle = 1,$$

$$\langle q^{2k}\alpha|q^{2l}\alpha\rangle = e^{|\alpha|^2(q^{2(l-k)}-1)},\ 0 \le k, l \le n-1.$$

To calculate orthogonality and normalization conditions we apply the following lemma; For $q^{2n} = 1$, $0 \le s \le n-1$,

- $1 + q^{2m} + q^{4m} + \cdots + q^{2m(n-1)} = n\delta_{m,0(mod\, n)}$
- $1 + q^{2(m-s)} + q^{4(m-s)} + \cdots + q^{2(m-s)(n-1)} = n\delta_{m,s(mod\, n)}$

where

$$\delta_{m,s\,(mod\, n)} = \begin{cases} 1, & m = s\ (mod\, n); \\ 0, & m \neq s\ (mod\, n). \end{cases} \tag{26}$$

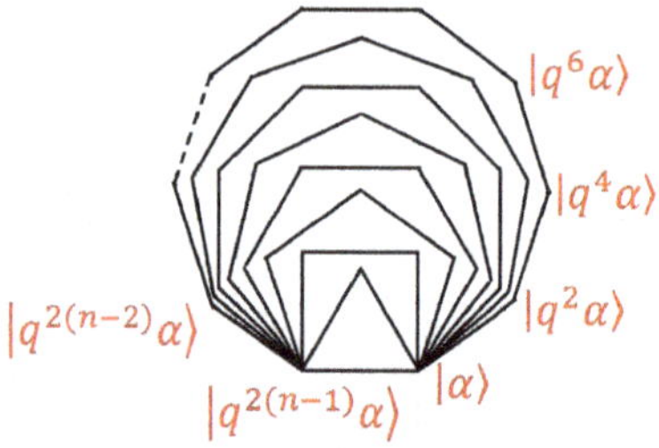

Fig. 4 The regular n-polygon

$$\begin{bmatrix} |0\rangle_\alpha \\ |1\rangle_\alpha \\ |2\rangle_\alpha \\ |3\rangle_\alpha \\ \vdots \\ |n-1\rangle_\alpha \end{bmatrix} = \mathbf{NQ} \begin{bmatrix} |\alpha\rangle \\ |q^2\alpha\rangle \\ |q^4\alpha\rangle \\ |q^6\alpha\rangle \\ \vdots \\ |q^{2(n-1)}\alpha\rangle \end{bmatrix} \tag{27}$$

Fig. 5 General structure of kaleidoscope states

5.1 *Quantum Fourier Transformation*

Our construction (Fig. 5) shows that orthogonal kaleidoscope of coherent states can by described by the Quantum Fourier transform

$$\begin{bmatrix} |\widetilde{0}\rangle_\alpha \\ |\widetilde{1}\rangle_\alpha \\ |\widetilde{2}\rangle_\alpha \\ |\widetilde{3}\rangle_\alpha \\ \vdots \\ |\widetilde{n-1}\rangle_\alpha \end{bmatrix} = \frac{1}{\sqrt{n}} \begin{bmatrix} 1 & 1 & 1 & \dots & 1 \\ 1 & w & w^2 & \dots & w^{n-1} \\ 1 & w^2 & w^4 & \dots & w^{2(n-1)} \\ 1 & w^3 & w^6 & \dots & w^{3(n-1)} \\ \vdots & \vdots & \vdots & \ddots & \vdots \\ 1 & w^{(n-1)} & w^{2(n-1)} & \dots & w^{(n-1)(n-1)} \end{bmatrix} \begin{bmatrix} |\alpha\rangle \\ |q^2\alpha\rangle \\ |q^4\alpha\rangle \\ |q^6\alpha\rangle \\ \vdots \\ |q^{2(n-1)}\alpha\rangle \end{bmatrix}, \tag{28}$$

where $w = e^{\frac{-2\pi i}{n}} = \bar{q}^2$ is the nth root of unity, so that corresponding transformation matrix, the Vandermonde matrix as generalized Hadamard gate,

$$|\widetilde{k}\rangle_\alpha = \frac{1}{\sqrt{n}} \sum_{j=0}^{n-1} w^{jk} |q^{2j}\alpha\rangle \quad 0 \le k \le n-1, \tag{29}$$

is the unitary gate $QQ^\dagger = Q^\dagger Q = I$. For orthonormal states we define normalization matrix,

$$\mathbf{N} = \frac{e^{\frac{|\alpha|^2}{2}}}{\sqrt{n}} \operatorname{diag}\left({}_0e^{|\alpha|^2}, {}_1e^{|\alpha|^2}, {}_2e^{|\alpha|^2}, \ldots, {}_{n-1}e^{|\alpha|^2}\right)^{-1/2} (mod\, n)$$

in terms of $(mod\ n)$ exponential functions:

$$f_s(|\alpha|^2) =_s e^{|\alpha|^2} (mod\, n) \equiv \sum_{k=0}^{\infty} \frac{(|\alpha|^2)^{nk+s}}{(nk+s)!}, \quad 0 \le s \le n-1. \tag{30}$$

These functions represent superposition of standard exponentials

$${}_se^{|\alpha|^2} (mod\, n) = \frac{1}{n} \sum_{k=0}^{n-1} \bar{q}^{2sk} e^{q^{2k}|\alpha|^2}, \quad 0 \le s \le n-1, \tag{31}$$

related to each other by derivatives

$$\frac{\partial}{\partial |\alpha|^2}\left[{}_se^{|\alpha|^2}\right] = {}_{s-1}e^{|\alpha|^2}, \quad \frac{\partial}{\partial |\alpha|^2}\left[{}_0e^{|\alpha|^2}\right] =_{n-1} e^{|\alpha|^2}.$$

According to this, function f_s defined in (30) is a solution of ordinary differential equation of degree n

$$f_s^{(n)} = f_s, \quad \text{where } 0 \le s \le n-1, \tag{32}$$

with proper initial values: $f_s^{(s)}(0) = 1$ and

$$f_s(0) = f_s'(0) = \cdots = f_s^{(s-1)}(0) = f_s^{(s+1)}(0) = \cdots = f_s^{(n-1)}(0) = 0.$$

As we have learned recently, these functions as the generalized hyperbolic functions were introduced also in [13]. By using these functions one can derive compact expression for the kaleidoscope states as

$$|\alpha\rangle = e^{-\frac{1}{2}|\alpha|^2} e^{\alpha \hat{a}^\dagger} |0\rangle \Rightarrow |s\rangle_\alpha = \frac{{}_se^{\alpha \hat{a}^\dagger}}{\sqrt{{}_se^{|\alpha|^2}}} |0\rangle \quad (mod\, n), \quad 0 \le s \le n-1. \tag{33}$$

5.2 *Number of Photons in Kaleidoscope of Quantum Coherent States*

Cyclic permutation of kaleidoscope states, generated by annihilation operator $\hat{a}$, allows us to calculate average number of photons in these states

$$\hat{a}|s\rangle_\alpha = \alpha \frac{N_s}{N_{s-1}} |s-1\rangle_\alpha \quad \Rightarrow \tag{34}$$

$$_\alpha\langle s|\widehat{N}|s\rangle_\alpha = |\alpha|^2 \left[\frac{_{s-1}e^{|\alpha|^2}}{_s e^{|\alpha|^2}} \right], 1 < s \leq n-1, \tag{35}$$

$$\hat{a}|0\rangle_\alpha = \alpha \frac{N_0}{N_{n-1}} |n-1\rangle_\alpha \quad \Rightarrow \tag{36}$$

$$_\alpha\langle 0|\widehat{N}|0\rangle_\alpha = |\alpha|^2 \left[\frac{_{n-1}e^{|\alpha|^2}}{_0 e^{|\alpha|^2}} \right]. \tag{37}$$

Asymptotically they approach the coherent states average number value

$$\lim_{|\alpha|\to\infty} {_\alpha\langle s|\widehat{N}|s\rangle_\alpha} \approx |\alpha|^2 = \langle\alpha|\widehat{N}|\alpha\rangle$$

while for small occupation numbers give integers

$$\lim_{|\alpha|\to 0} {_\alpha\langle s|\widehat{N}|s\rangle_\alpha} = s .$$

6 Quantum Algebra

Our kaleidoscope coherent states (33) are eigenstates of operator $q^{2\hat{N}}$:

$$q^{2\hat{N}}|k\rangle_\alpha = q^{2k}|k\rangle_\alpha, \quad k = 0, 1 \ldots, n-1.$$

In the Fock space this operator is an infinite matrix of the form

$$\Sigma_3 \equiv q^{2\hat{N}} = I \otimes \begin{pmatrix} 1 & 0 & 0 & \ldots & 0 \\ 0 & q^2 & 0 & \ldots & 0 \\ 0 & 0 & q^4 & \ldots & 0 \\ \vdots & \vdots & \vdots & \ddots & \vdots \\ 0 & 0 & 0 & \ldots & q^{2(n-1)} \end{pmatrix}, \quad \Sigma_1 = I \otimes \begin{pmatrix} 0 & 0 & 0 & \ldots & 1 \\ 1 & 0 & 0 & \ldots & 0 \\ 0 & 1 & 0 & \ldots & 0 \\ \vdots & \vdots & \vdots & \ddots & \vdots \\ 0 & 0 & 0 & \ldots & 0 \end{pmatrix}. \tag{38}$$

Here the $n \times n$ matrices are called the Sylvester clock and shift matrices correspondingly. They are q-commutative

$$\Sigma_1 \Sigma_3 = q^2 \Sigma_3 \Sigma_1,$$

satisfy relations

$$\Sigma_1^n = I, \quad \Sigma_3^n = I$$

and are connected by the unitary transformation:

$$\Sigma_1 = (I \otimes Q) q^{2\hat{N}} (I \otimes Q^+).$$

Hermann Weyl in book [14] proposed them for description of quantum mechanics of finite dimensional systems. By dilatation operator $q^{2\hat{N}}$ we define q^2-number operator

$$[\hat{N}]_{q^2} = \frac{q^{2\hat{N}} - 1}{q^2 - 1}$$

for non-symmetrical q-calculus, and

$$[\hat{N}]_{\tilde{q}^2} = \frac{q^{2\hat{N}} - q^{-2\hat{N}}}{q^2 - q^{-2}}$$

for the symmetrical one. In our kaleidoscope basis, these number operators are diagonal and given by q-numbers:

$$[\hat{N}]_{q^2} = diag([0]_{q^2}, [1]_{q^2}, \ldots, [n-1]_{q^2}),$$

with $[n]_{q^2} = \frac{q^{2n}-1}{q^2-1}$ for non-symmetric case, and

$$[\hat{N}]_{\tilde{q}^2} = diag([0]_{\tilde{q}^2}, [1]_{\tilde{q}^2}, \ldots, [n-1]_{\tilde{q}^2}),$$

with $[n]_{\tilde{q}^2} = \frac{q^{2n}-q^{-2n}}{q^2-q^{-2}}$ for the symmetrical one.

For symmetric case the q-number operator is Hermitian and can be factorized as

$$[\hat{N}] = \hat{B}^+ \hat{B}, \quad [\hat{N} + 1] = \hat{B} \hat{B}^+,$$

where

$$\hat{B} = \hat{a} \sqrt{\frac{[N]_{\tilde{q}^2}}{N}}.$$

Explicitly in matrix form it is

$$\hat{B} = I \otimes \begin{pmatrix} 0 & \sqrt{[1]} & 0 & \ldots & 0 \\ 0 & 0 & \sqrt{[2]} & \ldots & 0 \\ \vdots & \vdots & \vdots & \ddots & \vdots \\ 0 & 0 & 0 & \ldots & 0 \end{pmatrix}, \quad \hat{B}^+ = I \otimes \begin{pmatrix} 0 & 0 & 0 & \ldots & 0 \\ \sqrt{[1]} & 0 & 0 & \ldots & 0 \\ 0 & \sqrt{[2]} & 0 & \ldots & 0 \\ \vdots & \vdots & \vdots & \ddots & \vdots \\ 0 & 0 & 0 & \ldots & 0 \end{pmatrix} \tag{39}$$

and $\hat{B}^n = 0$, $(\hat{B}^+)^n = 0$. In non-symmetric case the number operator is not Hermitian.

6.1 Symmetric Case

For symmetric case we have the quantum algebra

$$\hat{B}\hat{B}^{+} - q^{2}\hat{B}^{+}\hat{B} = q^{-2\hat{N}}, \tag{40}$$

$$\hat{B}\hat{B}^{+} - q^{-2}\hat{B}^{+}\hat{B} = q^{2\hat{N}}, \tag{41}$$

and quantum q^2-oscillator with Hamiltonian

$$\hat{H} = \frac{\hbar\omega}{2}\left([\hat{N}]_{\tilde{q}^2} + [\hat{N} + I]_{\tilde{q}^2}\right).$$

In the kaleidoscope states as the eigenstates, the spectrum of this Hamiltonian is

$$E_k = \frac{\hbar\omega}{2}\frac{\sin\frac{2\pi}{n}(k+\frac{1}{2})}{\sin\frac{\pi}{n}}. \tag{42}$$

The same spectrum was obtained in [15] for description of physical system of two anyons. Appearance of quantum algebraic structure in two different physical systems, as optical coherent states and the anyons problem is instructive.

6.2 Non-symmetric Case

In this case the quantum algebra of operators is q^2-deformed

$$\hat{B}\hat{B}^{+} - q^{2}\hat{B}^{+}\hat{B} = I, \tag{43}$$

$$\hat{B}\hat{B}^{+} - \hat{B}^{+}\hat{B} = q^{2\hat{N}}, \tag{44}$$

with periodic (mod n) ($[k+n]_{q^2} = [k]_{q^2}$) q^2-numbers

$$[k]_{q^2} = e^{i\frac{\pi}{n}(k-1)}\frac{\sin\frac{\pi}{n}k}{\sin\frac{\pi}{n}}. \tag{45}$$

7 Conclusions

Kaleidoscope of coherent states considered in present paper can be realized by proper phase superposition of coherent states of light (the Gaussian states) and it can provide a unit of quantum information corresponding not only to diadic, but also to an arbitrary number base n. These states furnish the representation of quantum symmetry related with quantum q-oscillator.

As a generalization of the Schrödinger cat states, from our kaleidoscope states one can construct multi qudits entangled quantum states. This work is in progress.

Acknowledgements This work is supported by TUBITAK grant 116F206.

References

1. Pashaev, O.K.: J. Phys. Conf. Ser. **482**, 012033 (2014)
2. Pashaev, O.K.: J. Phys. Conf. Ser. **766**, 012015 (2016)
3. Pashaev, O.K.: Phys. Scr. **90**, 074010 (2015)
4. Vitiello, G.: Phys. Lett. A **376**, 2527–2532
5. Bialynicka-Birula, Z.: Phys. Rev. **173**, 172 (1968)
6. Stoler, D.: Phys. Rev. D **4**, 2309 (1971)
7. Spiridonov, V.: Phys. Rev. A **52**, 1909 (1995)
8. Dodonov, V.V., Malkin, I.A., Man'ko, V.I.: Physica **72**, 597 (1974)
9. Schrödinger, E.: Naturwissenschaften **14**, 664 (1926)
10. Wolfgang, P.S.: Quantum Optics in Phase Space. Wiley, London (2001)
11. Ralph, T.C., Gilchrist, A., Milburn ,G.J., Munro ,W.J., Glancy, S.: Phys. Rev. A **68**(4) (2003)
12. Peremolov, A.M.: Generalized Coherent States and Their Applications. Text and Monographs in Physics. Springer, Berlin (1986)
13. Ungar, A.: Am. Math. Mon. **98**, 688 (1982)
14. Weyl, H.: The Theory of Groups and Quantum Mechanics. Dover, New York (1931)
15. Floratos, E.G., Tomaras, T.N.: Phys. Lett. B **251**, 163–166 (1990)

CPSIA information can be obtained
at www.ICGtesting.com
Printed in the USA
BVHW010317081118
532517BV00001B/3/P

9 783030 013752